Miklos Bodanszky

# Peptide Chemistry

## A Practical Textbook

With 2 Figures and 5 Tables

Springer-Verlag Berlin Heidelberg New York
London Paris Tokyo

Professor Dr. Miklos Bodanszky
One Markham Road 1 E
Princeton, NJ 08540, USA

ISBN 3-540-18984-X Springer-Verlag Berlin Heidelberg New York
ISBN 0-387-18984-X Springer-Verlag New York Berlin Heidelberg

Library of Congress Cataloging-in-Publication Data
Bodanszky, Miklos. Peptide chemistry. Bibliography: p. Includes index. 1. Peptides. I. Title
QD431.B75 1988   547.7′56   88-19985
ISBN 0-387-18984-X

© Springer-Verlag Berlin Heidelberg 1988
Printed in the United States of America

The use of registered names, trademarks, etc. in the publication does not imply, even in the absence of a specific statement, that such names are exempt from the relevant protective laws and regulations and therefore free for general use.

Product Liability: The publishers can give no guarantee for information about drug dosage and application thereof contained in this book. In every individual case the respective user must check its accuracy by consulting other pharmaceutical literature.

Typesetting: Konrad Triltsch, Würzburg
2151/3145-543210

*To the memory of
my brother Dr. S. Bodanszky*

# Preface

Nature applied peptides for a great variety of specific functions. The specificity provided by the individual character of each amino acid is further ehanced by the combination of several amino acids into larger molecules. Peptides therefore can act as chemical messengers, neurotransmitters, as highly specific stimulators and inhibitors, regulating various life-processes. Entire classes of biologically active compounds, such as the opioid peptides or the gastrointestinal hormones emerged within short periods of time and it is unlikely that the rapid succession of discoveries of important new peptides would come to a sudden halt. In fact, our knowledge of the field is probably still in an early stage of development. Peptides also gained importance in our everyday life. A dipeptide derivative is used as sweetener, another to control high blood pressure. Peptides of moderate size, such as oxytocin or vasopressin have been available for medical purposes for many years and more recently peptide hormones of considerable chain-length, for instance calcitonin and secretin, are manufactured for therapy. These practical aspects alone might be considered as justification for teaching peptide chemistry in a special course. Yet, not less important is the experience that the chemistry of peptides is treated as an orphan in textbooks both of organic chemistry and of biochemistry, a topic disposed of in a few pages. There are good reasons for this. While biosynthesis and metabolism of amino acids and biosynthesis and function of proteins are in the main line of biochemistry, peptide chemistry, and the synthesis of peptides particularly, are too strongly tied to organic chemistry to be treated in detail in biochemistry texts. A more than casual presentation of peptides, however, would certainly transcend the limits of single-volume organic chemistry textbooks. Also, sequence determination lies outside the general methods of structure determination applied for the majority of organic compounds and synthesis of peptides aims at compounds with molecular weights well beyond those encountered in the synthesis of other natural products, with the notable exception of nucleic acids. The ensuing problems, like simultaneous handling of numerous functional groups, the continued need for conservation of chiral purity and unusual modes of retrosynthetic analysis are particular to peptides. Peptide chemistry is indeed a discipline in itself with its own practical significance and intellectual challenge. It deserves to be taught in a special course.

The author of this small volume has taught Peptide Chemistry as a graduate course through many years. A suitable textbook was sorely missed both by students and instructor and the copious handouts were insufficient substitutes. This textbook was written with the assumption that its existence might stimulate the teaching of its subject. The size of the volume was limited by the desire to make it inexpensive, but the material presented was intended to be adequate for a one-semester three-credit course. The first part, dealing with the determination of peptide structure, could be used separately in a short course while the second part, on synthesis, might support a one-semester two-credit course. It would give the author considerable satisfaction if instructors, prompted by the availability of this small textbook, would offer a course on Peptide Chemistry and if the students would find it helpful in their studies.

The comments and suggestions by Professor G. T. Young (Oxford University) concerning my book Principles of Peptide Synthesis (Springer Verlag, 1984) were very helpful in writing the present volume and are gratefully acknowledged. I thank my wife Agnes for numerous corrections and improvements in the preparation of this manuscript.

Princeton, May 1988                                    M. Bodanszky

# Table of Contents

**Part Two: Peptide Synthesis**

# I. Introduction

The importance of proteins, substances entrusted with primary functions in the living cell, need not be stressed any more. We may pay tribute to the nomenclator who from the Greek *protos* (first) or *proteios* (primary) coined the word *protein*. This foresight or intuition, usually attributed to Jac. Berzelius, remains vindicated in spite of the most impressive progress in nucleic acid chemistry and the emergence of DNA as the carrier of genetic information and RNA as template in protein biosynthesis. Nucleic acids provide the blueprint for the construction of complex machinery, the machines themselves are proteins. In fact, information encoded in DNA-s and in RNA-s is operative only in the presence of enzymes, that is proteins. It is clear therefore, that protein chemistry is one of the most important chapters of biochemistry, and can even stand in itself as the subject of a textbook. It is perhaps less obvious why *peptide chemistry* should be treated separately. The term "peptide" (from pepsis = digestion or peptones = digestion products of proteins) denotes relatively small compounds which are quite similar to proteins except that latter are substances of higher molecular weight. The reasons for this distinction are not self evident. There is no distinct borderline between the two groups of materials; molecules built of 100 or more amino acid residues are usually regarded as proteins and those containing a lesser number of residues as peptides. Chemistry remains essentially the same when the number of amino acid residues in a compound increases or decreases. Furthermore, a major part of peptides originates from proteins which are cleaved by specific enzymes at "signals" incorporated into their chains. These would be good reasons to discuss peptides in the framework of protein chemistry but at least equally strong arguments suggest that a textbook should be dedicated solely to peptide chemistry.

Protein chemistry is a vast topic. Its many sides encompass the description of various kinds of proteins, fibrous and globular molecules, enzymes, oxygen carriers and electrontransfer agents, glucoproteins, lipoproteins, phosphoproteins, carriers of thyroxin, of cholesterol, hormones, hormone precursors and other regulators including their physical aspects, such as molecular weight, aggregation and subunit structure. Last but not least it deals with the biosynthesis of proteins. Thus, protein chemistry is clearly a large subdiscipline of biochemistry. In contrast, peptide chemistry, because of the smaller molecular weight of the compounds that belong to its territory, is closer to organic chemistry. Also, a separate class of peptides, mostly cyclic, contains unusual amino acids, which are not constituents of proteins. The possibility to synthesize peptides through the means of organic chemistry seems to be sufficient justification for the separate treatment that is attempted in the following chapters.

Amino acids, the constituents of both peptides and proteins are a broad subject in themselves, too broad to be included into this volume. The interested reader should turn to the three-volume treatise by J. P. Greenstein and M. Winitz (published in 1961 but not yet outdated) or to the more recent work of Barrett (1983). Here we restrict the discussion to nomenclature and to general methods of synthesis. Table 1 shows the twenty amino acids of which proteins are made. Only for these constituents is information encoded in DNA and RNA. Certain amino acid derivatives, such as hydroxyproline, hydroxylysine, phosphate and sulfate esters of serine and tyrosine are also found in proteins, but these are produced via post-translational changes on the already assembled chains through the intervention of specific enzymes. In addition to their trivial names, Table 1 lists the self-explanatory three-letter abbreviations generally accepted for them and also the less easily memorized single letter abbreviations which are, however, quite practical in the comparisons of analogs and in the demonstration of homology between peptides. Table 2 serves to illustrate the much larger class of amino acids found in microbial peptides.

Synthesis of amino acids is a task seldom undertaken by peptide chemists. The commonly occurring amino acids are commercially available: they are manufactured on large industrial scale often via fermentation processes. From time to time, however, an amino acid is needed that is not listed in research-supply catalogs. For instance a study of the role of the amino group in the side chain of a lysine residue requires an analog in which this amino group is replaced by a methyl group. The corresponding amino acid, 2-aminoheptanoic acid or homonorleucine must be prepared for this purpose. A general method of amino acid synthesis based on an inexpensive starting material, acetyl-laminomalonic acid diethyl ester, can be applied in the majority of such instances. Alkylation of the sodio derivative is followed by saponification of the ester groups with alkali and then by hydrolysis with acid, which is accompanied by decarboxylation:

$$
\begin{array}{ccc}
& \text{COOEt} & \\
& | & \\
\text{CH}_3\text{CO}-\text{NH}-\text{C}-\text{H} & \xrightarrow{\text{NaOEt}} & \text{CH}_3\text{CO}-\text{NH}-\text{C}-\text{Na} \xrightarrow{\text{R}-\text{Br}} \\
& | & \\
& \text{COOEt} &
\end{array}
$$

$$
\begin{array}{ccc}
\text{COOEt} & & \text{COOEt} \\
| & & | \\
\text{CH}_3\text{CO}-\text{NH}-\text{C}-\text{R} & \xrightarrow{\text{OH}^-} & \text{CH}_3\text{CO}-\text{NH}-\text{C}-\text{R} \xrightarrow{\text{H}^+/\text{HOH}} \\
| & & | \\
\text{COOEt} & & \text{COO}^-
\end{array}
$$

$$
\text{CO}_2 + \text{Cl}^- \cdot \overset{+}{\text{H}}_3\text{N}-\text{CHR}-\text{COOH}
$$

**Table 1.** Amino acid constituents of proteins

General structure of L-amino acids (in Fischer projection):

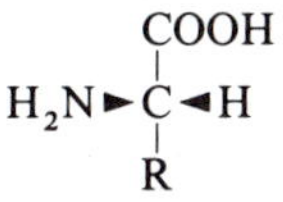

$$\begin{array}{c} COOH \\ | \\ H_2N \blacktriangleright C \blacktriangleleft H \\ | \\ R \end{array}$$

| | name | side chain R | abbre-viation | single letter code |
|---|---|---|---|---|
| monoamino mono-carboxylic acids | glycine | $H-$ | Gly | G |
| | alanine | $CH_2-$ | Ala | A |
| | valine | $(CH_3)_2CH-$ | Val | V |
| | leucine | $(CH_3)_2CHCH_2-$ | Leu | L |
| | isoleucine | $CH_3CH_2(CH_3)CH-$ | Ile | I |
| acidic amino acids and their amides | aspartic acid | $HOOC-CH_2-$ | Asp | D |
| | asparagine | $H_2NCO-CH_2-$ | Asn | N |
| | glutamic acid | $HOOC-CH_2CH_2-$ | Glu | E |
| | glutamine | $H_2NCO-CH_2CH_2-$ | Gln | Q |
| basic amino acids | lysine | $H_2NCH_2CH_2CH_2CH_2-$ | Lys | K |
| | arginine | $H_2NC(NH)NH-CH_2CH_2CH_2-$ | Arg | R |
| | histidine | 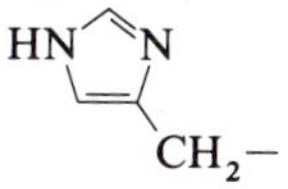 | His | H |
| hydroxy amino acids | serine | $HO-CH_2-$ | Ser | S |
| | threonine | $CH_3CH(OH)-$ | Thr | T |
| aromatic amino acids | phenylalanine | 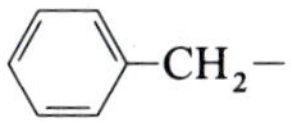 | Phe | F |
| | tyrosine | 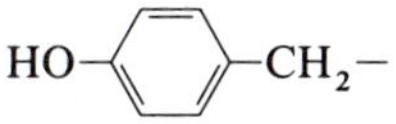 | Tyr | Y |
| | tryptophan | 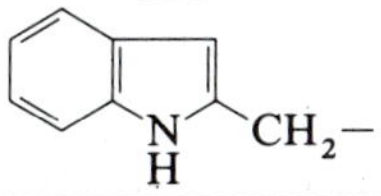 | Trp | W |
| sulfur contg. amino acids | cysteine | $HS-CH_2-$ | Cys | C |
| | methionine | $CH_3-S-CH_2CH_2-$ | Met | M |
| an imino acid | proline | 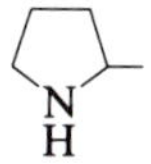 | Pro | P |

**Table 2.** Some amino acid constituents of microbial peptides (amino acids not found in proteins)

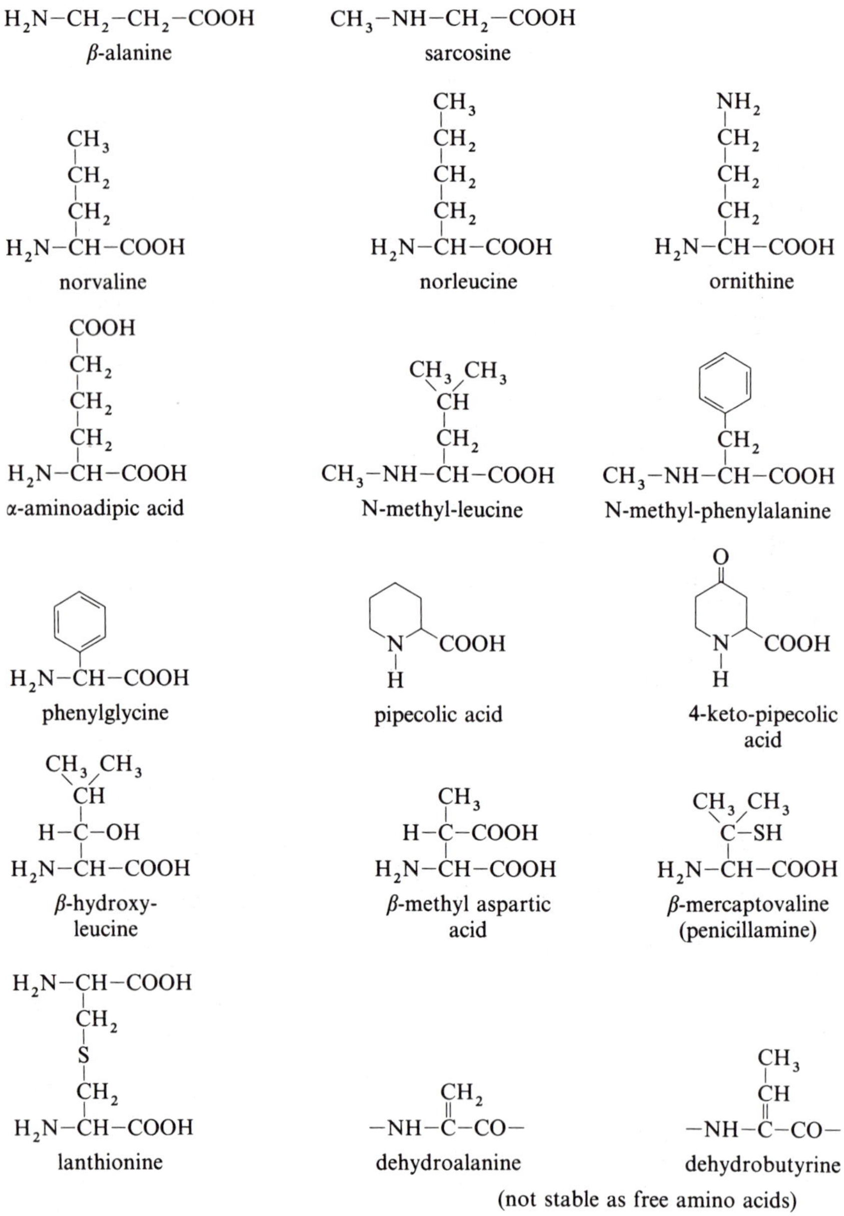

H$_2$N–CH$_2$–CH$_2$–COOH
β-alanine

CH$_3$–NH–CH$_2$–COOH
sarcosine

$$
\begin{array}{c}
\text{CH}_3 \\
|\\
\text{CH}_2 \\
|\\
\text{CH}_2 \\
|\\
\text{H}_2\text{N–CH–COOH}
\end{array}
$$
norvaline

$$
\begin{array}{c}
\text{CH}_3 \\
|\\
\text{CH}_2 \\
|\\
\text{CH}_2 \\
|\\
\text{CH}_2 \\
|\\
\text{H}_2\text{N–CH–COOH}
\end{array}
$$
norleucine

$$
\begin{array}{c}
\text{NH}_2 \\
|\\
\text{CH}_2 \\
|\\
\text{CH}_2 \\
|\\
\text{CH}_2 \\
|\\
\text{H}_2\text{N–CH–COOH}
\end{array}
$$
ornithine

$$
\begin{array}{c}
\text{COOH} \\
|\\
\text{CH}_2 \\
|\\
\text{CH}_2 \\
|\\
\text{CH}_2 \\
|\\
\text{H}_2\text{N–CH–COOH}
\end{array}
$$
α-aminoadipic acid

$$
\begin{array}{c}
\text{CH}_3\;\text{CH}_3 \\
\diagdown\diagup\\
\text{CH} \\
|\\
\text{CH}_2 \\
|\\
\text{CH}_3\text{–NH–CH–COOH}
\end{array}
$$
N-methyl-leucine

$$
\begin{array}{c}
\bigcirc \\
|\\
\text{CH}_2 \\
|\\
\text{CH}_3\text{–NH–CH–COOH}
\end{array}
$$
N-methyl-phenylalanine

$$
\begin{array}{c}
\bigcirc \\
|\\
\text{H}_2\text{N–CH–COOH}
\end{array}
$$
phenylglycine

pipecolic acid

4-keto-pipecolic acid

$$
\begin{array}{c}
\text{CH}_3\;\text{CH}_3 \\
\diagdown\diagup\\
\text{CH} \\
|\\
\text{H–C–OH} \\
|\\
\text{H}_2\text{N–CH–COOH}
\end{array}
$$
β-hydroxy-leucine

$$
\begin{array}{c}
\text{CH}_3 \\
|\\
\text{H–C–COOH} \\
|\\
\text{H}_2\text{N–CH–COOH}
\end{array}
$$
β-methyl aspartic acid

$$
\begin{array}{c}
\text{CH}_3\;\text{CH}_3 \\
\diagdown\diagup\\
\text{C–SH} \\
|\\
\text{H}_2\text{N–CH–COOH}
\end{array}
$$
β-mercaptovaline (penicillamine)

$$
\begin{array}{c}
\text{H}_2\text{N–CH–COOH} \\
|\\
\text{CH}_2 \\
|\\
\text{S} \\
|\\
\text{CH}_2 \\
|\\
\text{H}_2\text{N–CH–COOH}
\end{array}
$$
lanthionine

$$
\begin{array}{c}
\text{CH}_2 \\
||\\
\text{–NH–C–CO–}
\end{array}
$$
dehydroalanine

$$
\begin{array}{c}
\text{CH}_3 \\
|\\
\text{CH} \\
||\\
\text{–NH–C–CO–}
\end{array}
$$
dehydrobutyrine

(not stable as free amino acids)

The hydrochloride thus obtained is converted to the zwitter-ionic form by neutralization with lithium hydroxide or with pyridine and dilution with ethanol. The byproduct, LiCl or pyridinium chloride remains in solution, while the free amino acid, as a dipolar ion, separates, mostly in crystalline form:

$$Cl^- \cdot H_3\overset{+}{N}-CHR-COOH \xrightarrow[\text{pyridine}]{\text{LiOH or}} H_3\overset{+}{N}-CHR-COO^-$$

Alternatively the hydrochloride of the target amino acid is passed through a column of a strong cation-exchange resin (in H-cycle) and eluted with dilute ammonia.

Needless to say that the product of synthesis is a racemate, from which the L and D enanthiomers have to be separated. Of the numerous methods of resolution a fairly general approach, enzyme catalyzed hydrolysis of acetyl-amino acids should be given special consideration. Acylase greatly enhances the rate of removal by hydrolysis of the acetyl group from L-amino acids, but leaves the (negligible) hydrolysis rate of the acetyl-D-amino acid unaffected. Hence, after completion of the hydrolysis, concentration of the neutral solution and dilution with absolute ethanol yields the L-amino acid, a dipolar ion insoluble in alcohol, while the salt of the unchanged acetyl derivative of the D-enanthiomer remains in solution.

$$CH_3CO-NH-CHR-COONa \xrightarrow[\text{acylase}]{\text{HOH}}$$

$$\underset{L}{H_3\overset{+}{N}-CHR-COO^-} + \underset{D}{CH_3CO-NH-CHR-COONa}$$

DL

There are considerable differences in the hydrolysis rates of different amino acids. If the rate is too low for practical purposes, then the chloroacetyl derivatives of the racemates can be applied as substrates instead of the acetyl derivatives. Of course, it is often worthwhile to recover the unchanged D-acylamino acid and hydrolyze it with aqueous acid to produce the D-enanthiomer of the amino acid. The "unnatural" D isomers are frequently used as building components in studies of structure-activity relationships, in the preparation of hormone analogs resistant to the action of proteolytic enzymes and in the synthesis of microbial peptides.

Amino acids were isolated from natural sources, asparagine from plants, cystine from urinary stones in the first decade of the last century and became known later to be constituents of proteins. The *peptide bond,* however, as the fundamental feature of protein structure was recognized only in this century (Hofmeister 1902; E. Fischer 1906). Several more decades elapsed until sequence elucidation reached the point of practicality. The turning point was signaled by the determination of the structure of insulin (Sanger et al. 1953). The subsequent development of automated amino acid analysis (Spackman, Stein and Moore 1958) and automatic sequencing (Edman and Begg 1967)

**Table 3.** Biologically active peptides

---

tuftsin:    Thr—Lys—Pro—Arg
rigin:      Gly—Gln—Pro—Arg    (immunoactive peptides)

bradykinin:    Arg—Pro—Pro—Gly—Phe—Ser—Pro—Phe—Arg
kallidin:      Lys—Arg—Pro—Pro—Gly—Phe—Ser—Pro—Phe—Arg

substance P:  Arg—Pro—Lys—Pro—Gln—Gln—Phe—Phe—Gly—Leu—Met—$NH_2$

angiotensin (human) and its precursors:

Asp—Arg—Val—Tyr—Ile—His—Pro—Phe—His—Leu—Val—Ile—His—Asn
renin substrate

⟶ (renin)

Asp—Arg—Val—Tyr—Ile—His—Pro—Phe—His—Leu + Val—Ile—His—Asn
angiotensin I

⟶ (converting enzyme)

Asp—Arg—Val—Tyr—Ile—His—Pro—Phe + His—Leu
angiotensin II

opioid peptides:
Leu enkephaline    Tyr—Gly—Gly—Phe—Leu
Met enkephaline    Tyr—Gly—Gly—Phe—Met
  β-endorphin (human):  Tyr—Gly—Gly—Phe—Met—Thr—Ser—Glu—Lys—Ser—
                        Gln—Thr—Pro—Leu—Val—Thr—Leu—Phe—Lys—Asn—
                        Ala—Ile—Ile—Lys—Asn—Ala—Tyr—Lys—Lys—Gly—Glu

melanocyte stimulating hormone (α-MSH):  Ac—Ser—Tyr—Ser—Met—Glu—His—
                        Phe—Arg—Trp—Gly—Lys—Pro—Val—$NH_2$

oxytocin:
Cys—Tyr—Ile⌐
|              |
Cys—Asn—Gln↩
|
Pro—Leu—Gly—$NH_2$

arginine-vasopressin:
Cys—Tyr—Phe⌐
|              |
Cys—Asn—Gln↩
|
Pro—Arg—Gly—$NH_2$

calcitonin (human):  Cys—Gly—Asn—Leu—Ser—Thr—Cys—Met—Leu—Gly—
                     Thr—Thr—Thr—Gln—Asp—Phe—Asn—Lys—Phe—His—
                     Thr—Phe—Pro—Gln—Thr—Ala—Ile—Gly—Val—Gly—Ala—
                     Pro—$NH_2$

corticotropin:  Ser—Tyr—Ser—Met—Glu—His—Phe—Arg—Trp—Gly—Lys—Pro—
                Val—Gly—Lys—Lys—Arg—Arg—Pro—Val—Lys—Val—Tyr—Pro—
                Asn—Gly—Ala—Glu—Asp—Glu—Ser—Ala—Glu—Ala—Phe—
                Pro—Leu—Glu—Phe

gastrin (human):  Pyr—Glu—Gly—Pro—Trp—Leu—Glu—Glu—Glu—Glu—Glu—
                  Ala—Tyr—Gly—Trp—Met—Asp—Phe—$NH_2$
                  (Pyr = pyroglutamyl)

---

**Table 3.** (continued)

glucagon and secretin:
    (both porcine)
    His − Ser − Gln − Gly − Thr − Phe − Thr − Ser − Asp − Tyr − Ser − Lys − Tyr − Leu −
    His − Ser − Asp − Gly − Thr − Phe − Thr − Ser − Glu − Leu − Ser − Arg − Leu − Arg −

    Asp − Ser − Arg − Arg − Ala − Gln − Asp − Phe − Val − Gln − Trp − Leu − Met − Asn − Thr
    Asp − Ser − Ala − Arg − Leu − Gln − Arg − Leu − Leu − Gln − Gly − Leu − Val − NH₂

insulin (human):

    G I V E Q C C T S I C S L Y Q L E N Y C N
              \                 /
    F V N Q H L C G S H L V E A L Y L V C G E R G F F Y T P K T

insulin (porcine):

    G I V E Q C C T S I C S L Y Q L E N Y C N
              \                 /
    F V N Q H L C G S H L V E A L Y L V C G E R G F F Y T P K A

insulin (bovine):

    G I V E Q C C A S V C S L Y Q L E N Y C N
              \                 /
    F V N Q H L C G S H L V E A L Y L V C G E R G F F Y T P K A

reduced the determination of the primary structure of peptides and proteins to a mere routine and only the establishment of the three-dimensional structure (conformation, geometry, architecture) is regarded by now as a true objective of research. The methodology of peptide synthesis first progressed faster than structure determination. The earliest modes of peptide bond formation pioneered by Curtius (1881) and Fischer (1902) at the turn of the century yielded impressive but not yet practical results. Introduction of the amino-protecting benzyloxycarbonyl group (Bergmann and Zervas 1932) led to a new era of peptide synthesis. Improvements in the methods of peptide bond formation, particularly the invention of carbonic acid mixed anhydrides (Wieland and Bernhard 1951; Boissonnas 1951) gave new impetus to peptide synthesis and in 1953 the methodology reached a degree of sophistication which allowed the synthesis of a peptide hormone, oxytocin, by V. du Vigneaud and his associates. From here on synthesis of peptides progressed by leaps and bounds. Introduction of dicyclohexylcarbodiimide, a still unsurpassed coupling reagent (Sheehan and Hess 1955) had a major impact on the methodology of peptide bond formation and further refinement was brought about by the development of active esters (Schwyzer 1953; Bodanszky 1955). At least equally important improvements could be noted in the methods of protection: acid labile blocking groups built on the stability and thus ready formation of the tert.butyl cation (Carpino 1957), the tert.butyloxycarbonyl group particularly, remain among

**Table 4.** Microbial peptides

| | |
|---|---|
| gramicidin S: | 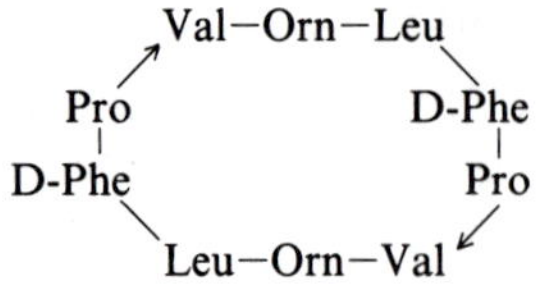 |
| etamycin: | 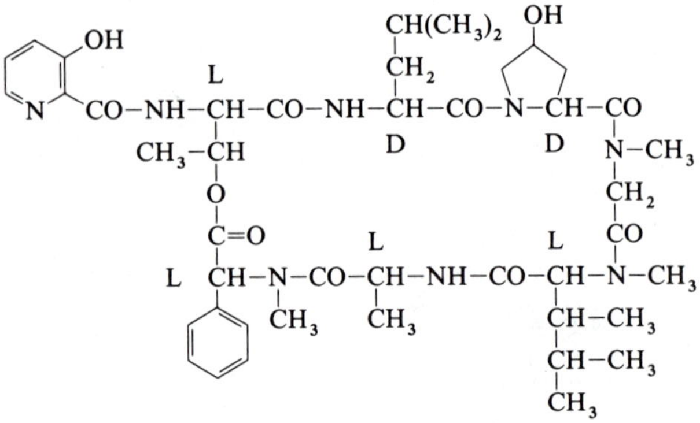 |
| mikamycin B | 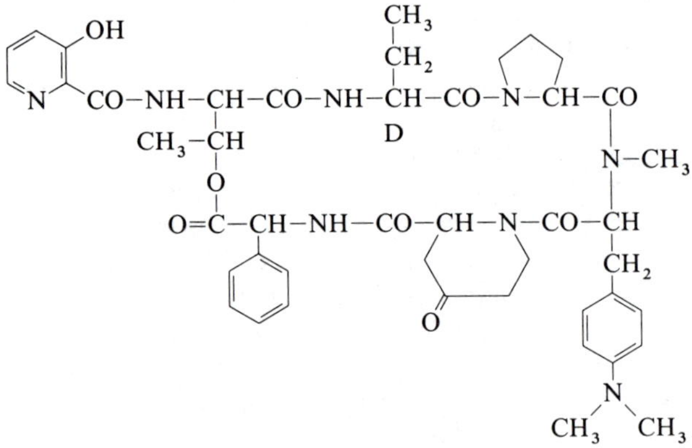 |
| 6-aminopenicillanic acid:<br>(6-APA) | 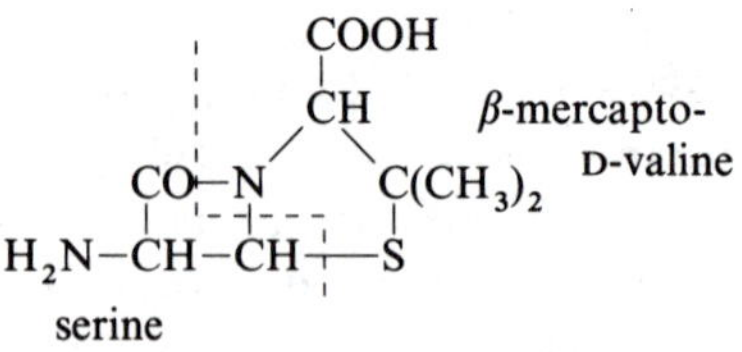 |
| penicillin G: | 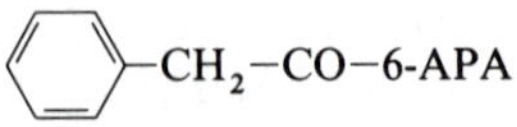 |
| ampicillin:<br>(semisynthetic penicillin) | |

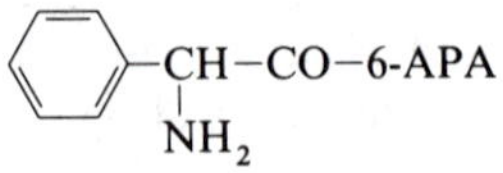

the tools of unchallenged importance even after the introduction of base sensitive blocking in the form of the 9-fluorenylmethyloxycarbonyl (Fmoc) group (Carpino and Han 1972). Yet, the most conspicuous milestone in the history of peptide synthesis is the invention of solid phase peptide synthesis by R. B. Merrifield (1963). Not lastly because of the automation made possible by his technique a large number of recently discovered biologically active peptides became available soon after their isolation and sequence determination. In order to illustrate the rapid growth of the field, a selection of such compounds is listed in Table 3. In Table 4 some microbial peptides are drawn. A mere inspection of their structures reveals that synthesis of peptides remains a challenge for the organic chemist. Yet, even the preparation of peptides containing only the usual amino acid constituents is complicated by undesired side reactions. These are so numerous that major review articles were written about them (e.g. Bodanszky and Martinez 1983). Therefore, it seems to be unwise to embark on an important task in peptide synthesis without a substantial knowledge of the chemistry involved in the process. The following chapters should provide the groundwork for students who seek a degree of competence in peptide chemistry.

# References

Bodanszky, M.: Nature *175*, 685 (1955)

Bodanszky, M., Martinez, J.: In The Peptides, vol. 5 (Gross, E., Meienhofer, J. eds.) p. 111, New York, Academic Press 1983

Boissonnas, R. A.: Helv. Chim. Acta *34*, 874 (1951)

Carpino, L. A.: J. Amer. Chem. Soc. *79*, 4427 (1957)

Carpino, L. A.: Han, G. Y.: J. Org. Chem. *37*, 3404 (1972)

Curtius, T.: J. Prakt. Chem. *24*, 239 (1881)

du Vigneaud, V., Ressler, C., Swan, J. M., Roberts, C. W., Katsoyannis P. G., Gordon, S.: J. Amer. Chem. Soc. *75*, 4879 (1953)

Edman, P., Begg, G.: Eur. J. Biochem. *1*, 80 (1967)

Fischer, E.: Ber. dtsch. Chem. Ges. *35*, 1095 (1902)

Fischer, E.: Ber. dtsch. Chem. Ges. *39*, 530 (1906)

Hofmeister, F.: Ergeb. Physiol. Chem. Pharmacol. *1*, 759 (1902)

Merrifield, R. B.: J. Amer. Chem. Soc. *85*, 2149 (1963)

Sanger, F.: Nature *171*, 1025 (1953)

Schwyzer, R.: Helv. Chim. Acta *36*, 414 (1953)

Sheehan, J. C., Hess, G. P.: J. Amer. Chem. Soc. *77*, 1067 (1955)

Spackman, D. H., Stein, W. H., Moore, S.: Anal. Chem. *30*, 1190 (1958)

Wieland, T., Bernhard, H.: Liebigs Ann. Chem. *572*, 190 (1951)

# Additional Sources

Barrett, G. C.: Chemistry and Biochemistry of Amino Acids. Chapman & Hall, London, 1983

Greenstein, J. P., Winitz, M.: Chemistry of the Amino Acids. Wiley & Sons, New York, 1961

Roberts, D. C., Vellacio, F.: Unusual Amino Acids in Peptide Synthesis, in The Peptides, vol. 5, Gross, E., Meienhofer, J. eds. Academic Press, New York, 1983

# Part One: Structure Determination

# II. Amino Acid Analysis

A most revealing information that can readily be obtained about a peptide is its amino acid composition. Yet, the results of amino acid analysis are really meaningful only if the sample consists of a single peptide. Analysis of mixtures is usually an unrewarding effort. Thus, purification should precede analysis and this generalization is valid for most other methods of structure determination as well. Homogeneity as a prerequisite of analysis can not be overemphasized. Purification is sometimes possible simply by crystallization but in most cases chromatography, electrophoresis or countercurrent distribution or a combination of these techniques is needed.

One can identify and quantitatively determine the amino acid constituents of a peptide by non-destructive methods, for instance by nuclear magnetic resonance spectroscopy, but in almost every instance peptides are *hydrolyzed* and their amino acid composition determined through the analysis of the hydrolysate. Hydrolytic cleavage of the peptide bond does not consume energy yet it does not occur spontaneously: an energy barrier has to be overcome by catalysis. Proteolytic enzymes are capable to effect complete hydrolysis of peptides at room temperature, in neutral solution and usually within hours. Nevertheless, only in special cases are enzymes used for the purpose of amino acid analysis. Such a case is the detection of racemization during peptide synthesis (cf. chapter VIII). In the general practice a small sample of the peptide is hydrolyzed with constant boiling (ca. 5.7 N) hydrochloric acid at 110 °C for 16 – 24 hours. Hydrolysis time varies from laboratory to laboratory and is also a function of the peptide in question. Peptide bonds between hindered residues such as valine or isoleucine are unusually resistant to hydrolysis and require 48 or more hours for complete cleavage. Unfortunately, serine and threonine are gradually destroyed in the process. The thus arising conflict can be resolved by parallel analyses with shorter and longer periods of hydrolysis. The exact values for serine and threonine are established by graphical extrapolation to zero time while the corrected values for the hindered amino acids are obtained by extrapolation for a very long hydrolysis time. It is somewhat more difficult to deal with the decomposition of tryptophan. It is possible to counteract it by the use of highly purified hydrochloric acid which contains no metallic elements or by replacing HCl with methanesulfonic acid and particularly by using a solution of 2-mercaptoethanesulfonic acid for hydrolysis. Exclusion of air is generally helpful: best results are obtained when hydrolysis is carried out in evacuated and sealed ampoules. The tryptophan problem can also be solved through the recording of the ultraviolet absorption spectrum of the intact peptide. The high molecular extinction coefficient of the indole system permits an exact determination of tryptophan content and this remains possible even in tyrosine containing peptides because on addition of alkali the absorption maximum of the phenolic ring shifts toward the visible while the maximum of tryptophan remains virtually unchanged.

  The actual determination of amino acids in the hydrolysate was once based on esterification and separation of the esters by fractional distillation. This time and material consuming approach became more practical after the advent of vapor phase chromatography. The necessary volatility is achieved by trimethylsilylation

$$CH_3-C\underset{NSiMe_3}{\overset{OSiMe_3}{\diagup}} + H_2N-CHR-COOH \longrightarrow$$

$$Me_3SiNH-CHR-COOSiMe_3 + CH_3CONH_2$$

or by trifluoroacetylation followed by esterification, e.g. with 1-butanol

$$H_2N-CHR-COOH \xrightarrow{(CF_3CO)_2O} CF_3CO-NH-CHR-COOH \xrightarrow[HCl]{1-BuOH}$$

$$CF_3CO-NH-CHR-CO-OCH_2CH_2CH_2CH_3$$

The general practice however, relies today on separation of the amino acids themselves by ion-exchange chromatography or by partition chromatography. This latter is performed on special columns under high pressure (high pressure liquid chromatography, HPLC). Automatic instruments permit the execution of a complete analysis within a short time, often less than one hour.

  A principal question, the monitoring of the effluent from the column remains to be discussed here. Small amounts of the individual amino acids emerging in the sequence of their elution have to be revealed and quantitated. In earlier procedures these amounts were in the range of 10 to 100 nanomoles but with improvements in the methodology much smaller samples can now be applied. The classical color reagent for the detection of nanomole quantities is ninhydrin, which reacts with amino acids in a transamination-decarboxylation reaction to yield Ruhemann's purple:

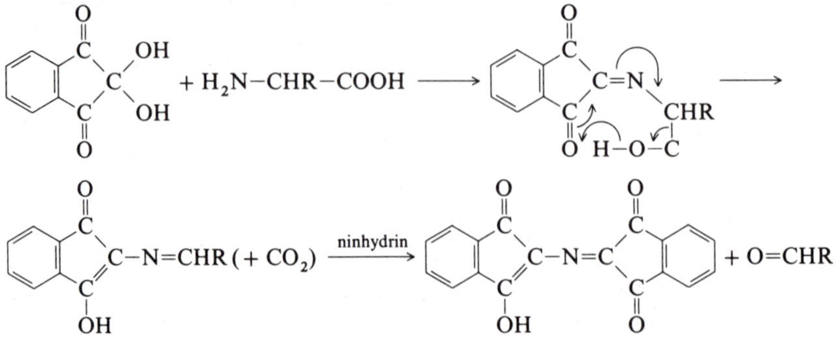

Absorption at 560 nm is used to determine the amount of the amino acid eluted from the column. Proline, a secondary amine, gives a yellow colored compound and this necessitates a second monitoring wavelength (440 nm). Continuous recording of absorption results in an elution diagram which is then compared with a standard curve for evaluation. Contemporary instruments provide a printout of the computed results of the analysis.

Further decrease in the amounts sufficient for amino acid analysis became possible by changing from the determination of extinction to the measurement of the intensity of light emitted by fluorescence. An extremely sensitive method is based on fluorescamine and its reaction with primary amines:

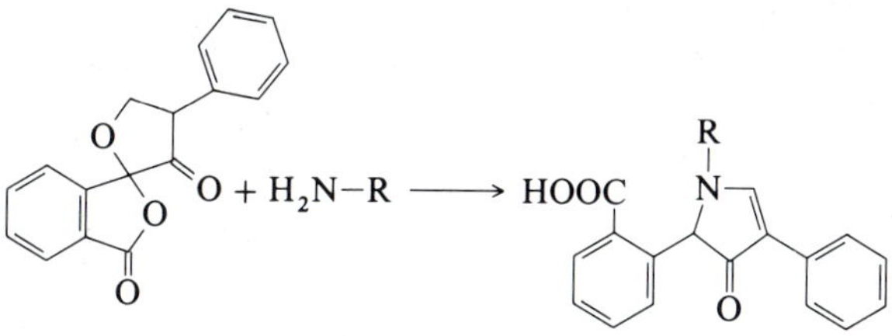

The fluorescence thus generated allows the determination of picomole quantities of amino acids. A somewhat less sensitive but also less expensive fluorescence reagent is o-phtalaldehyde

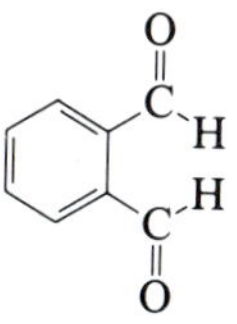

which, in the presence of mercaptoethanol, detects minute amounts of primary amines, but does not react with proline.

In connection with the application of highly sensitive methods of detection and hence very small amounts of peptides used for amino acid analysis a general difficulty must be pointed out. Our environment is full of trace amounts of amino acids. If nothing else, our fingerprints contaminate the glassware and other pieces of equipment in the laboratory: glycine, serine and aspartic acid are particularly ubiquitous and can confuse the results of amino acid analyses. Therefore, when small samples and sensitive methods are used, extreme care should be given to the cleaning of glassware and to the purity of the reagents used in the process.

## Additional Sources

Benson, J. R., Louie, P. C., Bradshaw, R. A.: Amino Acid Analysis of Peptides, in The Peptides, Vol. 4 (Gross, E., Meienhofer, J., eds.) pp. 217–260, New York: Academic Press 1981

Brenner, M., Niederwieser, A.: Thin Layer Chromatography of Amino Acids in Methods of Enzymology, Vol. 11 (Hirs, C. H. ed.) pp. 39–59, New York: Academic Press 1967

# III. Sequence Determination

The order or *sequence* of individual amino acid residues along the peptide chain defines the covalent structure of the molecule. It is also called *primary structure* in order to make a clear distinction from the three dimensional geometry in peptides and proteins. The latter is generated by non-covalent forces such as hydrogen bonds between amide groups (*secondary structure*) and combination of polar and non-polar interactions and of disulfide bridges which result in chain folding (*tertiary structure*).

## A. End Group Determination

At one end of a peptide chain an amino acid residue is present with a free, that is unacylated amino group. This residue is called the amino terminal or more commonly N-terminal residue. At the opposite end of the chain a residue with a free carboxyl group is the carboxyl terminal or C-terminal residue:

$$H_2N-CHR^1-CO-NH-CHR^2-CO-\cdots\cdots-NH-CHR^{n-1}-CO-NH-CHR^n-COOH$$
N-terminal residue                                              C-terminal residue

This nomenclature is applied in a broad sense, mainly to indicate the direction of the peptide chain. Thus, in peptides in which an N-terminal glutamine residue is cyclized

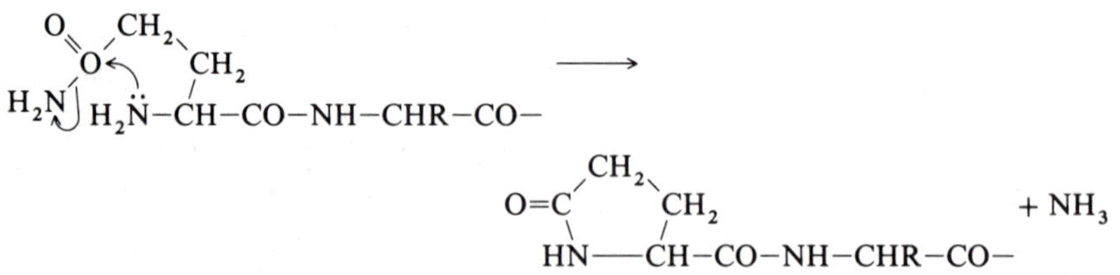

the pyroglutamyl residue is considered as N-terminal in spite of the absence of a free amino group. In an analogous manner the N-terminal amino acid residue retains this designation even if its amino group is acylated by a non-amino acid moiety, such as an acetyl group. Similarly, in peptide amides the C-terminal residue has no free carboxyl group and yet it is called C-terminal. End group analysis discussed in the next sections is possible however only when the terminal residues have indeed a free amino group or a free carboxyl group respectively.

The significance of end group analysis diminished with the development of methods of sequential degradation, which are treated in some detail later in this chapter. In earlier years determination of end groups formed an integral part of sequence analysis. Large peptides were cleaved into smaller fragments by partial acid hydrolysis, the fragments separated by chromatography and examined. The structure of a dipeptide fragment directly followed from its amino acid composition and from the determination of one of the end groups. For tripeptides both end groups had to be establsihed. The information collected from the analysis of the fragments was then assembled in the manner of jig-saw puzzle solving to establish the sequence of the starting material, the large peptide. This approach served admirably in the elucidation of structure of insulin, oxytocin and the vasopressins but it is more or less abandoned by now. Nevertheless, we present here a brief discussion of this subject, and not merely for historical reasons, but because these simple methods are still applied in special cases, for instance in the structure determination of small peptides with unusual amino acid constituents.

## 1. Determination of the N-terminal Residue

Substitution of the free amino group with an easily detectable group provides a "tag" for the N-terminal residue. The bond between the substituent and the amine nitrogen should remain intact during hydrolysis of peptide bonds in order to allow the isolation and identification of the tagged amino acid, the N-terminal residue. Nucleophilic displacement on aromatic nuclei is an obvious avenue toward this goal, but it requires activation with electron-withdrawing groups. For instance one of the nitro-groups in trinitrotoluene is readily displaced by the amino group of peptides

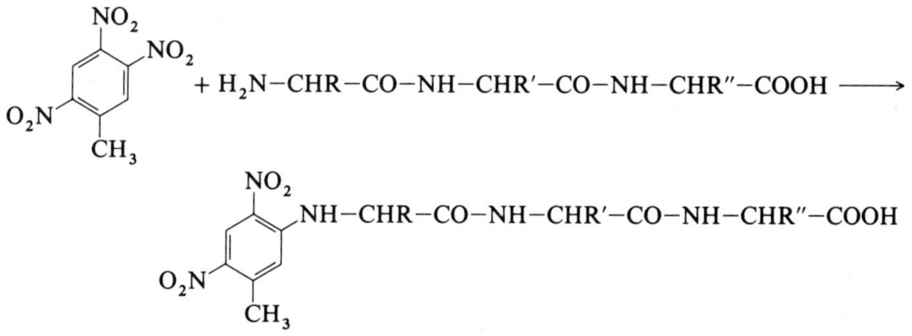

but even more suitable is the reagent found by Sanger (1945) the 2,4-dinitrofluorobenzene. In the presence of aqueous sodium bicarbonate it reacts at room temperature and converts the peptide to the 2,4-dinitrophenyl (DNP) derivative:

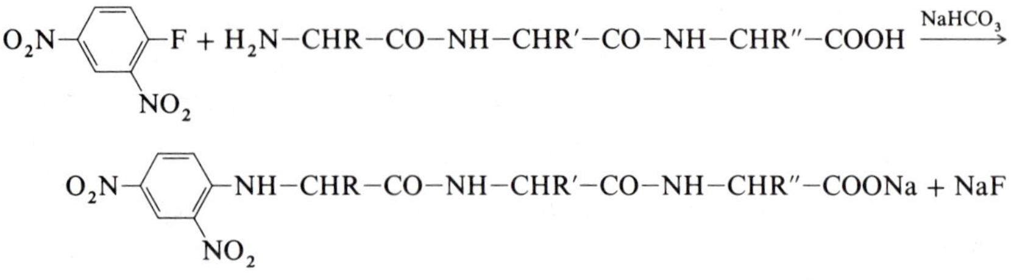

$$O_2N-\langle\text{ring}\rangle-F + H_2N-CHR-CO-NH-CHR'-CO-NH-CHR''-COOH \xrightarrow{NaHCO_3}$$

with $NO_2$ substituent on ring

$$O_2N-\langle\text{ring}\rangle-NH-CHR-CO-NH-CHR'-CO-NH-CHR''-COONa + NaF$$

with $NO_2$ substituent on ring

Hydrolysis of the DNP-peptide with constant boiling hydrochloric acid yields the constituent amino acids with the exception of the N-terminal one which is present as DNP-amino acid. After acidification this can be extracted with an organic solvent and identified by comparison with authentic samples of DNP-amino acids on chromatograms. The yellow color of DNP-amino acids facilitates their detection, but sensitivity is enhanced if the chromatograms are viewed under ultraviolet light. A more significant increase of sensitivity is necessary when the amount of peptide available for sequence analysis is very limited. Light emission by fluorescent groups opens up important possibilities. Thus, acylation of the free amino group of an N-terminal residue with 5-dimethylamino-naphtalene-1-sulfonic acid chloride ("dansyl chloride") provides a suitable tag (Gray and Hartley 1963), because the sulfonamide bond is much more resistant to acid hydrolysis than the peptide bond

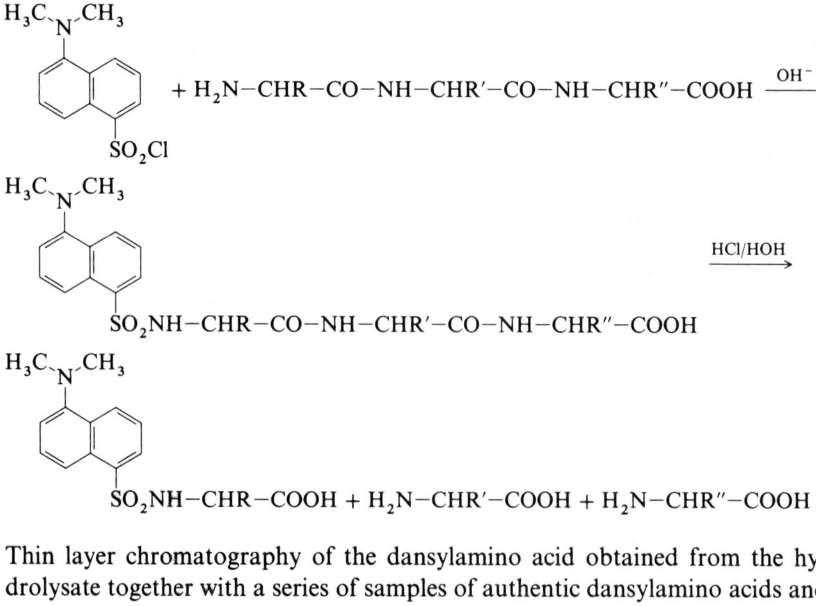

$H_3C\diagdown N\diagup CH_3$

[naphthalene ring structure with $SO_2Cl$]

$$+ H_2N-CHR-CO-NH-CHR'-CO-NH-CHR''-COOH \xrightarrow{OH^-}$$

$H_3C\diagdown N\diagup CH_3$

[naphthalene ring structure]

$$\xrightarrow{HCl/HOH}$$

$$SO_2NH-CHR-CO-NH-CHR'-CO-NH-CHR''-COOH$$

$H_3C\diagdown N\diagup CH_3$

[naphthalene ring structure]

$$SO_2NH-CHR-COOH + H_2N-CHR'-COOH + H_2N-CHR''-COOH$$

Thin layer chromatography of the dansylamino acid obtained from the hydrolysate together with a series of samples of authentic dansylamino acids and inspection of the fluorescent spots under ultraviolet light is a highly efficient method for the identification of the N-terminal residue in peptides.

Finally, the important principle of *subtractive analysis* should be mentioned. Selective decomposition of the N-terminal residue is possible, for instance by deamination with nitrous acid or through oxidative deamination with ninhydrin. Quantitative amino acid analysis before and after the reaction will show the absence of one amino acid in the hydrolysate of the treated sample. Obviously this is the residue with the free amino group hence it must have occupied the N-terminal position in the sequence.

## 2. Determination of the C-terminal Residue

Numerous methods were proposed for this purpose, but only few withstood the test of time. A reliable procedure is hydrazinolysis (Akabori et al. 1952) which involves the heating of a solution of the peptide in ca. 97% hydrazine in a sealed tube at 100 °C for 12 hours. The peptide bonds are cleaved by hydrazine and the amino acid constituents thus converted to amino acid hydrazides except the C-terminal residue which is merely liberated in the process. Its separation is facilitated by dinitrophenylation of the mixture with 2,4-dinitrofluorobenzene. The DNP-amino acid hydrazides as neutral substances are extracted from the aqueous, bicarbonate containing mixture with an organic solvent while the sodium salt of the DNP-derivative of the C-terminal amino acid remains dissolved:

$$H_2N-CHR-CO-NH-CHR'-CO-NH-CHR''-COOH \xrightarrow{H_2NNH_2}$$

$$H_2N-CHR-CONHNH_2 + H_2N-CHR'-CONHNH_2 + H_2N-CHR''-COO^-$$

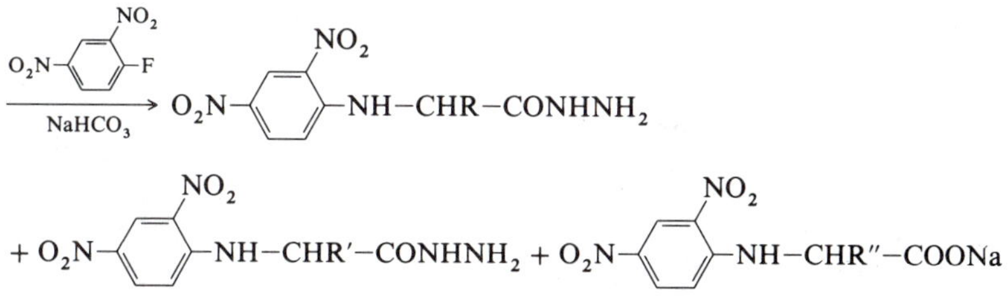

Of course, the DNP-derivatives of the dicarboxylic acid monohydrazides also remain in the aqueous layer.

Subtractive methods have been applied for the determination of the C-terminal residue in various ways. For instance, treatment of a peptide with a mixture of acetic anhydride and pyridine at elevated temperature (Dakin and

West 1928) converts the C-terminal residue to a methyl keton, presumably through a mixed anhydride and an azlactone

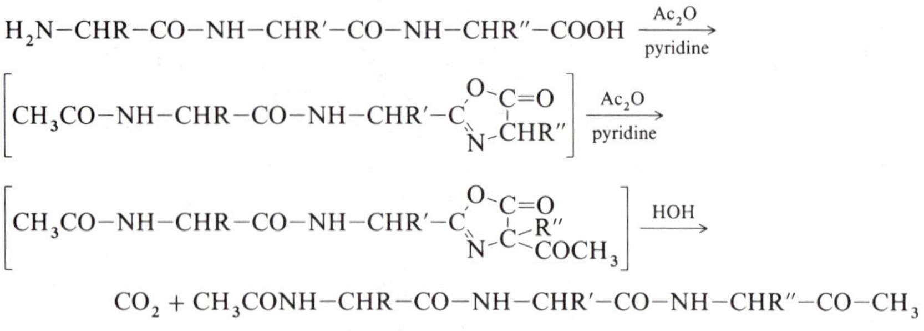

$$H_2N-CHR-CO-NH-CHR'-CO-NH-CHR''-COOH \xrightarrow[\text{pyridine}]{Ac_2O}$$

$$\left[ CH_3CO-NH-CHR-CO-NH-CHR'-C\underset{N-CHR''}{\overset{O-C=O}{\diagdown}} \right] \xrightarrow[\text{pyridine}]{Ac_2O}$$

$$\left[ CH_3CO-NH-CHR-CO-NH-CHR'-C\underset{N-C-COCH_3}{\overset{O-C=O}{\diagdown}}{}^{R''} \right] \xrightarrow{HOH}$$

$$CO_2 + CH_3CONH-CHR-CO-NH-CHR'-CO-NH-CHR''-CO-CH_3$$

Hydrolysis of samples of both starting material and product followed by amino acid analysis of the hydrolysates reveals the C-terminal residue by its absence from the hydrolysate of the treated peptide. In an analogous manner the C-terminal residue can be transformed to a non-amino acid moiety by reduction with $LiAlH_4$ or $LiBH_4$ (Fromageot et al. 1950). More reliable results are obtained if prior to reduction the carboxyl group is esterified with the help of diazomethane. In addition to subtraction the C-terminal residue can be identified in the form of the resulting aminoalcohol as well:

$$H_2N-CHR-CO-NH-CHR'-CO-NH-CHR''-COOH \xrightarrow{CH_2N_2}$$

$$H_2N-CHR-CO-NH-CHR'-CO-NH-CHR''-CO-OCH_3 \xrightarrow{LiAlH_4}$$

$$H_2N-CHR-CO-NH-CHR'-CO-NH-CHR''-CH_2OH \xrightarrow{HCl/HOH}$$

$$H_2N-CHR-COOH + H_2N-CHR'-COOH + H_2N-CHR''-CH_2OH$$

# B. Sequential Degradation

In short peptides determination of the two end-groups can be quite meaningful, but the information gained in this manner is of only modest value in the sequence analysis of longer chains. Methods, which allow stepwise removal of individual residues appeared more productive and accordingly attracted much attention. A considerable number of proposals can be found in the literature, but only a few reached the stage of practical application and, at the time of writing this book only the Edman degradation (1950) is widely applied. Nevertheless, one of the more or less abandoned approaches, gradual hydrolysis by specific proteolytic enzymes still deserves a brief discussion.

## 1. Hydrolysis Catalyzed by Exopeptidases

Aminopeptidases catalyze the hydrolysis of the peptide bond which follows the N-terminal residue (A). As a consequence, the second residue (B) becomes N-terminal and is cleaved off

$$A-B-C-D-E-F-G-H \xrightarrow[\text{(}-A\text{)}]{\text{aminopeptidase}} B-C-D-E-F-G-H$$

$$\xrightarrow[\text{(}-B\text{)}]{\text{aminopeptidase}} C-D-E-F-G-H \xrightarrow[\text{(}-C\text{)}]{\text{aminopeptidase}} D-E-F-G-H$$

$$\xrightarrow[\text{(}-D\text{)}]{\text{aminopeptidase}} E-F-G-H \xrightarrow[\text{(}-E\text{)}]{\text{aminopeptidase}} F-G-H \text{ and so on}$$

Thus, at least in principle, the entire chain is degraded in a stepwise manner. Since the free amino acids appear in the order of their position in the sequence it would seem possible to determine the sequence of a peptide by taking samples of the reaction mixture at regular time intervals and subject them to quantitative amino acid analysis. In the practical execution of the experiment, however, only the position of the first few residues can be established. This is due to variations in the hydrolysis rates related to individual amino acid residues. When L-leucine occupies the N-terminal position the bond connecting it to the next residue is so rapidly cleaved by a renal enzyme that it was misnamed "leucine amino peptidase" although it is not specific for leucine. It catalyzes the cleavage of N-terminal methionine to about the same extent, but removes proline at a rate orders of magnitude lower than the one observed with leucine or methionine. Glycine is cleaved even more slowly. Thus, sequential degradation does not follow a simple pattern and it comes to a virtual halt at proline and glycine residues. An enzyme known as aminopeptidase M shows a lesser variation in the hydrolysis rates of the bonds following individual amino acid residues and is, therefore, more suitable for the complete degradation of a peptide. Nevertheless its main use is not in sequence determination but rather in the examination of chiral purity of synthetic peptides. Since the enzyme does not catalyze the cleavage of the peptide bond between a D-residue and the next amino acid, complete digestibility proves that no racemization occurred during synthesis.

The limitations noted for aminopeptidases are true for carboxypeptidases as well. The pancreatic enzyme carboxypeptidase A shows low rates in the hydrolysis of peptides with a basic amino acid (arginine or lysine) as the C-terminal residue. Carboxypeptidase B is particularly effective when the same basic residues occupy the C-terminal position. The yeast enzyme carboxypeptidase Y is less specific and therefore more generally applicable, but probably still unsuited for the elucidation of a longer sequence.

## 2. Edman Degradation

**a) The Two-step Procedure.** Phenylisocyanate reacts with the free amino group of the N-terminal residue in peptides to yield a phenylcarbamoyl derivative

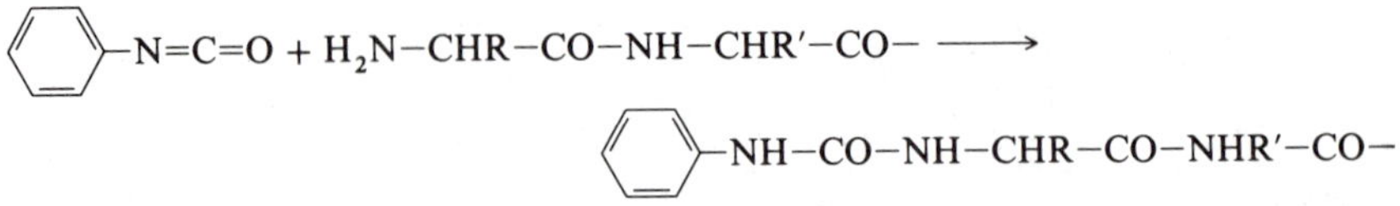

$$\text{〈⟩}-N{=}C{=}O + H_2N{-}CHR{-}CO{-}NH{-}CHR'{-}CO{-} \longrightarrow$$

$$\text{〈⟩}-NH{-}CO{-}NH{-}CHR{-}CO{-}NHR'{-}CO{-}$$

which, when treated with a strong acid such as HCl, affords a phenylhydantoin with the concomitant cleavage of the bond between the first and the second residue:

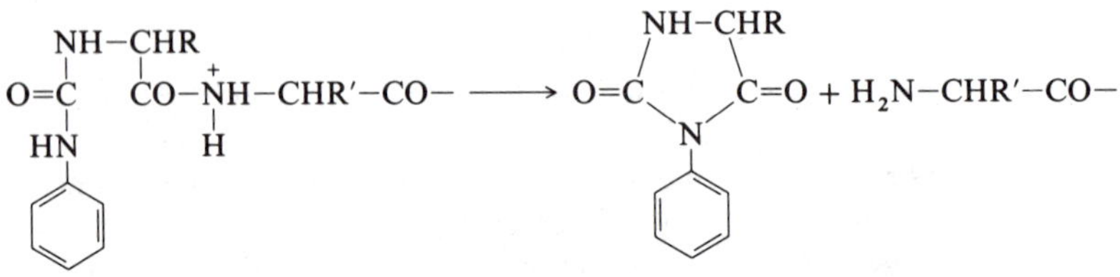

$$\begin{array}{c} NH{-}CHR \\ O{=}C \quad CO{-}\overset{+}{N}H{-}CHR'{-}CO{-} \\ HN \quad\quad H \end{array} \longrightarrow \begin{array}{c} NH{-}CHR \\ O{=}C \quad\quad C{=}O + H_2N{-}CHR'{-}CO{-} \\ N \end{array}$$

The new N-terminal residue can then be removed as a phenylhydantoin in a second cycle of phenylcarbamoylation and cyclization. This method of stepwise degradation, while discovered several decades earlier (Bergmann et al. 1927), reached practicality only in 1950 when Edman modified the reagent and applied chromatographic procedures for identification of the cyclic products. The improved reagent, phenylisothiocyanate, smoothly converts the peptide to the phenylthiocarbamoyl derivative which is cyclized and cleaved by the action of hydrochloric acid (dissolved in an organic solvent such as dioxane):

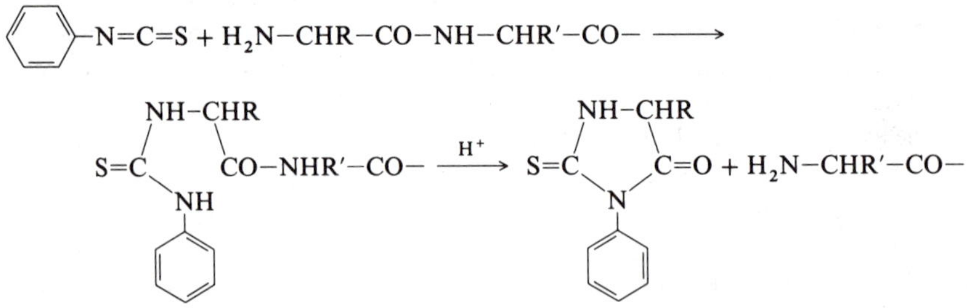

$$\text{〈⟩}-N{=}C{=}S + H_2N{-}CHR{-}CO{-}NH{-}CHR'{-}CO{-} \longrightarrow$$

$$\begin{array}{c} NH{-}CHR \\ S{=}C \quad CO{-}NHR'{-}CO{-} \\ NH \end{array} \xrightarrow{H^+} \begin{array}{c} NH{-}CHR \\ S{=}C \quad\quad C{=}O + H_2N{-}CHR'{-}CO{-} \\ N \end{array}$$

The phenylthiohydantoins formed in the process of *Edman degradation* were compared with authentic phenylthiohydantoins prepared from amino acids and thus identified. Yet, in spite of numerous further improvements, such as application of thin layer chromatography, vapor-phase chromatography and even mass spectrometry for the identification of phenylthiohydantoins, the method had serious limitations. Because of trace amounts of water retained in peptide samples, some hydrolysis of the chain takes place during cyclization with HCl. Fragments are generated with new N-termini and degraded simultaneously with the parent peptide. Therefore, after a few cycles the results become less and less clear and sequencing can not be further pursued.

**b) The Three-step Procedure.** A major breakthrough in sequence analysis was brought about by a thorough study of the mechanism of the Edman degradation that revealed that phenylthiohydantoin formation takes place in two distinct steps. First the sulfur atom in the phenylthiocarbamoyl peptide attacks the carbonyl carbon of the N-terminal residue and a thiazolidinone derivative is produced. This cyclization is the one that involves the cleavage of the peptide bond between the first two residues of the chain. The cycle is concluded with the acid catalyzed opening of the thiazolidinone and ring closure to a phenylthiohydantoin. Thus, the three steps of Edman degradtion can be formulated in the following way:

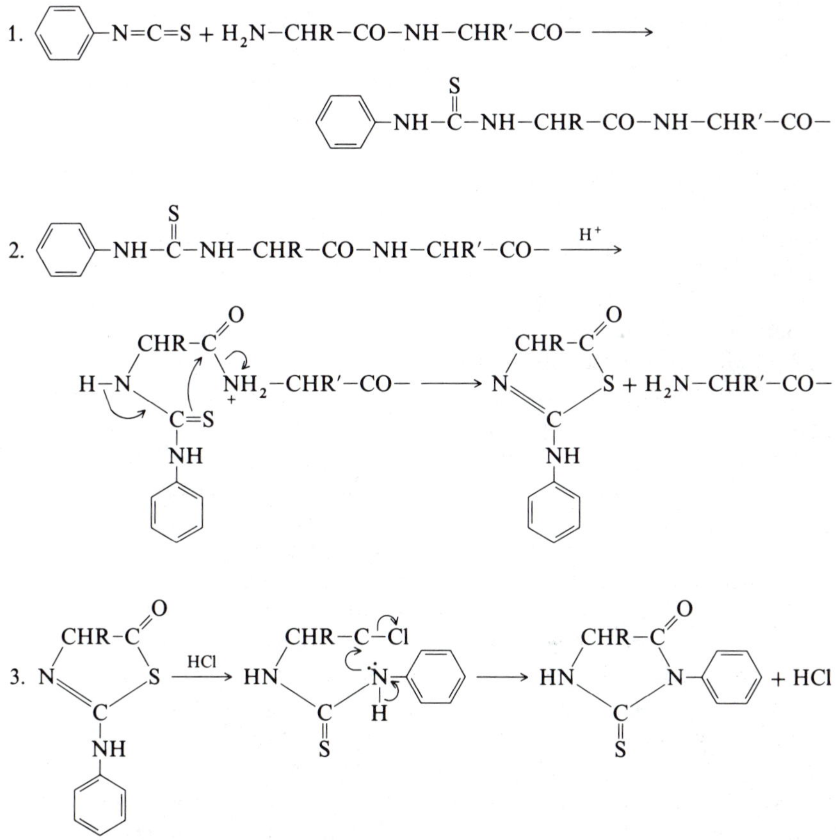

The second step, while acid catalyzed, does not require HCl. Strong organic acids, such as trifluoroacetic acid, are sufficient to induce thiazolidinone formation with selective bond cleavage and such acids are less prone to cause hydrolysis of the chain. Hence no unintentional formation of fragments occurs, no generation of misleading new N-terminal residues. It is true that HCl is needed for the third step, for the conversion of the phenylthiazolidinone to a phenylhydantoin, but this operation is carried out on the thiazolidinone already separated (by extraction with an organic solvent) from the shortened peptide and thus the latter is not exposed to the adverse effect of HCl. Accordingly the modified ("three step") Edman degradation can be pursued through many cycles. It would appear that the third step is not really necessary, but in the praxis of sequencing identification of the rather unstable thiazolidinones turned out not to be quite reliable. Phenylthiohydantoins are more suitable for this purpose. With the improved procedure determination of the sequence through forty or even fifty residues has been accomplished and the method gained almost exclusive application.

c) **Automated Sequencing.** In 1967 Edman and Begg described an instrument which performs the three step degradation through numerous cycles in an automated operation. By now several versions of the "sequenator" are commercially available, all built on the same principles but constructed according to modified methods of implementation. In the original "spinning cup" instrument the solution of a peptide forms a thin film on the wall of a small rapidly rotated glass container (the "cup"). The necessary reagents are syphoned in and out, the extracted thiazolidinones collected in a fraction collector and subsequently converted to the more stable phenylthiohydantoins for identification, mostly by high pressure liquid chromatography (HPCL). In the "solid phase" modification the peptide is attached through covalent bonding to an insoluble support, a polymer or a glass with a functional group (aminoglass). In the "gas-phase" (or more correctly vapor phase) processes the reagents are condensed on the sample and removed by evaporation. A series of additional improvements were introduced during the development of sequenators. For instance tetrakis-hydroxypropyl-propylenediamine or more recently tetrakis-hydroxyethyl-ethylenediamine is used to establish basic conditions for phenylthiocarbamoylation and pentafluoropropionic acid to provide sufficient acidity for cyclization of the phenylthiocarbamoyl derivative to the thiazolidinone. The latter are extracted with a mixture of dichloromethane and benzene. Selection of the solvent is guided by the desire to reduce losses during the sequencing procedure. Even a slight solubility of the peptide in organic solvents results in gradually decreasing recoveries and after a certain number of cycles sequencing becomes impractical. The vapor phase process represents a certain improvement in this respect. Of course volatile buffers are required; for instance dimethylbenzylamine can be used as base in phenylthiocarbamoylation. Losses can be further reduced by rendering the peptide more polar. An initial treatment with substituted phenylisothiocyanate can be applied

$$^-O_3S-\!\!\!\left\langle\!\!\!\bigcirc\!\!\!\right\rangle\!\!\!-N{=}C{=}S \;+\; \begin{matrix} NH_2 \\ | \\ CH_2 \\ | \\ CH_2 \\ | \\ CH_2 \\ | \\ CH_2 \\ | \\ -HN-CH-CO- \end{matrix} \;\longrightarrow\; \begin{matrix} S \\ \| \\ ^-O_3S-\!\!\!\left\langle\!\!\!\bigcirc\!\!\!\right\rangle\!\!\!-NH-C-NH \\ | \\ CH_2 \\ | \\ CH_2 \\ | \\ CH_2 \\ | \\ CH_2 \\ | \\ -HN-CH-CO- \end{matrix}$$

in order to modify the lysine side chains and to diminish thereby the solubility of the peptide in organic solvents.

Automatic sequencing with its numerous improvements is now routinely performed. A cycle requires an hour or less and the amount of peptide needed is well below the earlier applied 1 micromol. Yet, errors are not rare in the interpretation of results and the sequence of a peptide can be considered firmly established only if it is corroborated by a second, independent procedure.

**d) The Edman-Dansyl Process.** Sequence determination can also be carried out manually, without a sequenator and without the identification of phenylthiohydantoins. A simple approach is to perform quantitative amino acid analysis after each Edman cycle and identify each removed amino acid from the difference between two analyses. This is a reliable method but requires an amino acid analyzer and a peptide sample large enough to allow the taking of a whole series of samples for amino acid analyses. A less demanding procedure is based on end-group analysis by the dansyl method (cf. page 8). The N-terminal residue is identified by dansylation and hydrolysis of a small aliquot while the bulk of the peptide is subjected to an Edman cycle. The new N-terminal residue is determined, again on a small aliquot, and a second Edman cycle is carried out on the rest of the material and so on. All this is possible with commonly used laboratory equipment with identification of dansylamino acids on thin-layer plates.

# C. Sequence Determination with the Help of Mass Spectra

Sequencing a peptide by the recording of a single spectrum is a very attractive proposition which stimulated considerable research. The classical process of mass spectrometry requires a volatile sample which when injected into an evacuated chamber evaporates. The vapors are then exposed to a beam of electrons. Under electron bombardment the molecular ion and fragment ions are produced. For this purpose most peptides are not sufficiently volatile.

Substitution of polar groups yields a less polar and therefore more volatile material. Acetylation with acetic anhydride modifies the terminal amino group, the side chain amino group in lysine residues and also the hydroxyl groups in serine and threonine side chains in the desired sense. Volatility can be further increased by esterification of free carboxyls by treatment of the solution of the acetyl derivative in methanol with diazomethane:

$$H_2N-CHR-CO-NH-CHR'-CO-NH-CHR''-COOH \xrightarrow[2.\ CH_2N_2]{1.\ Ac_2O}$$

$$CH_3CO-NH-CHR-CO-NH-CHR'-CO-NH-CHR''-CO-OCH_3$$

The molecular ion of the derivatized peptide reveals only its molecular weight and has therefore limited importance. The charge propagates however along the chain and induces fragmentation. Fission of the C-terminal ester bond yields an ion with the mass of the molecular ion minus 31 (the methoxy group). This, usually abundant ion,

$$CH_3CO-NH-CHR-CO-NH-CHR'-CO-NH-CHR''-CO^+$$

would be in itself still insufficient to give information on the sequence, but further fragmentation occurs preferentially at the peptide bonds and the masses of the generated ions

$$CH_3CO-NH-CHR-CO-NH-CHR'-CO^+$$
$$CH_3CO-NH-CHR-CO^+$$
$$CH_3CO^+$$

provide the data sought in the experiment. The difference between two ions corresponds to the residue weight of the amino acid lost in the process and since these weights are characteristic for the individual amino acids it is possible to elucidate the sequence from the differences in the masses of a series of ions. A notable exception is the identical molecular weight of leucine and isoleucine: they can not be distinguished from each other in this simple manner. Unfortunately the actual picture is usually more complicated than the one sketched above. In addition to the principal fission at peptide bonds some secondary fragmentation takes place between the carbonyl groups and the adjacent α-carbon atom and side chains, particularly the bulkier ones, are also broken off from the peptide backbone. The confusion generated by such undesired fragmentation can be reduced by several methods, for instance by permethylation. The acetyl peptide methyl ester is treated with sodium hydride and then with methyl iodide

$$CH_3CO-NH-CHR-CO-NH-CHR'-CO-NH-CHR''-CO-OCH_3 \xrightarrow[2.\ CH_3I]{1.\ NaH}$$

$$CH_3CO-\underset{\underset{CH_3}{|}}{N}-CHR-CO-\underset{\underset{CH_3}{|}}{N}-CHR'-CO-\underset{\underset{CH_3}{|}}{N}-CHR''-CO-OCH_3$$

in order to further reduce polarity of the molecule and to increase its thermal stability. Of course, alkylation extends to other nitrogen atoms as well and the imidazole nucleus of histidine and the indole in tryptophan are also alkylated. While indeed simpler spectra can be secured on permethylated samples, some complications can arise from permethylation itself. For instance C-methylation of glycine residues occurs yielding alanine residues. The use of $CD_3I$ rather than $CH_3I$ allows the detection of unintentionally introduced methyl groups. An alternative improvement is achieved by dispensing with direct electron bombardment and applying "chemical ionization" instead. This means the transfer of a proton from ions such as $H_3^+$ or $CH_5^+$ to the molecule of the derivatized peptide. The masses of the molecular ion and of fragment ions are increased by one dalton and this has to be taken into consideration. There is less secondary fragmentation in this technique.

**Scheme 1.** Chemical ionization of a permethylated peptide and fragmentation of the peptide bond

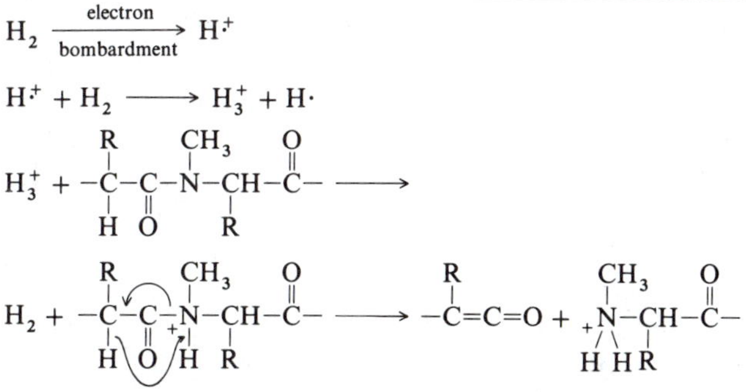

Further advancement in sequence determination by mass spectra was brought about through the introduction of new modes of ionization, such as field desorption and fast electron bombardment. Double focusing high resolution instruments and sophisticated computation greatly increased the number of peptides that could be sequenced by mass spectroscopy. For compounds which are unsuited for Edman degradation, such as cyclic peptides or peptides with an acyl group at their N-termini mass spectra can turn out to be the most practical avenue to sequence determination.

An interesting possibility is provided by a combination of enzymic degradation with mass spectrometry. Dipeptidyl-aminopeptidases and dipeptidyl-carboxypeptidases cleave two residues at a time which are removed in the form of dipeptides. Continuation of the process of catalyzed hydrolysis thus yields a series of dipeptides except, that in peptides with an odd number of residues the last amino acid appears as such. For instance digestion of the decapeptide

ABCDEFGHIJ with dipeptidyl-amino-peptidase gives rise to five dipeptides, AB, CD, EF, GH and IJ. A second sample of the same decapeptide is then subjected to a single cycle of Edman degradation and the resulting nonapeptide BCDEFGHIJ is hydrolyzed with dipeptidyl-aminopeptidase. Four dipeptides BC, DE, FG and HI are generated together with the amino acid J. Identity of the dipeptides and of the free amino acid can be established through mass spectroscopy, particularly with the chemical ionization technique, and one can even dispense with the separation of the fragments. The information secured in these experiments is sufficient for the determination of the sequence of the decapeptide.

# D. Fragmentation of Peptides

Since the word "peptide" usually denotes a relatively short chain of amino acid residues it is not obvious why peptides should be cut up to even smaller pieces prior to sequence determination. Fragmentation is indeed not always necessary. For instance mass spectrometry often can be used for sequencing of decapeptides or sometimes still larger molecules and Edman degradation can be pursued through numerous cycles. Yet, fragmentation might be desirable as a start of a second set of experiments which aim at corroboration of the findings of a first sequence determination carried out on the intact chain. Also, when no free amino group is present at the N-terminus, as is the case with cyclic peptides or with chains having an N-terminal pyroglutamyl residue or an acetylamino acid, treatment with phenylisothiocyanate becomes unproductive: it merely leads to the phenylthiocarbamoylation of side chain amino groups. In such instances fragmentation is clearly indicated, because it can generate new, more useful peptides which have free amino groups at their N-termini.

## 1. Hydrolysis with Specific Endopeptidases

Exopeptidases, enzymes which catalyze the hydrolytic cleavage of terminal residues have already been mentioned in this chapter. Here we wish to discuss the use of enzymes which cleave peptides midchain and which are selective enough to hydrolyze a peptide bond only when it follows certain amino acid residues. The enzyme most frequently applied for fragmentation is *trypsin*. Its usefulness is due only in part to the extreme rate enhancement of hydrolysis. More important is the high degree of specificity in its catalytic effect. Trypsin affects exclusively the bonds that follow the two basic amino acids, lysine and arginine. Sometimes also some cleavage after aromatic residues is observed but this is not the consequence of the practically negligible inherent chymotrypsin like activity of trypsin but rather of contamination of the enzyme preparation by chymotrypsin. The confusion created by such unintentional cleavage can be avoided if an irreversible chymotrypsin inhibitor such as the chloromethyl ketone related to and prepared from L-phenylalanine

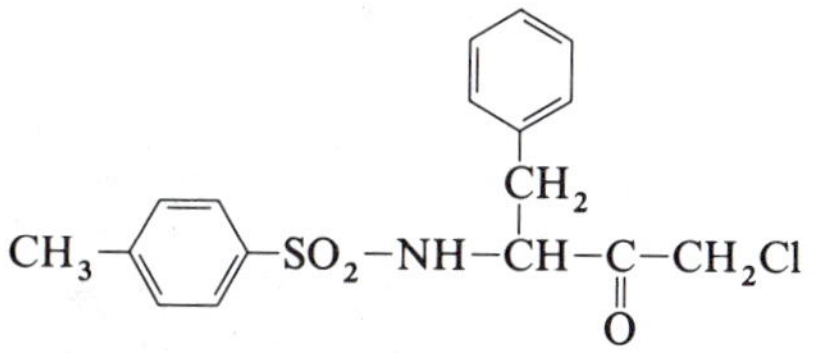

$$CH_3-\langle\bigcirc\rangle-SO_2-NH-\underset{\underset{\displaystyle CH_2}{|}}{CH}-\underset{\underset{\displaystyle O}{\|}}{C}-CH_2Cl$$

is added to the digestion mixture.

The rate of hydrolysis is affected by the amino acid that follows the basic residue in the sequence. Thus the Arg-Asp bond is cleaved more slowly than most other peptide bonds on the right side of arginine. Similarly, the bond between arginine and cysteic acid (the oxidized form of cysteine) is relatively resistant to tryptic hydrolysis. As it will be shown later such differences can be exploited in sequence studies, but it is possible to create even more substantial differences by appropriate modifications of the peptide molecule. For instance, treatment of a peptide that contains both lysine and arginine residues with maleic acid anhydride or preferably with citraconic acid anhydride yields derivatives in which the lysine residues have no free amino group in their side chain

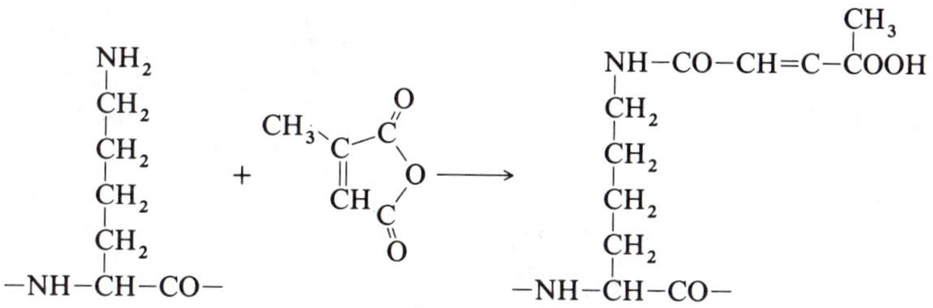

and hence are no more substrates for trypsin. In the maleyl or citraconyl derivative hydrolysis will occur exclusively at the arginine residues. The fragments can be separated and the maleyl or citraconyl groups removed under very mild conditions. Subsequent tryptic hydrolysis will yield a new set of fragments. Sequence determination is facilitated by the inherent information about the newly created C-terminal residues, namely arginine in the first set and lysine in the second set of tryptic fragments.

Trypsin can also be applied in cysteine containing peptides, because aminoethylation of the sulfhydryl group transforms the cysteine side chain to that of thialysine

$$
\begin{array}{ccc}
\text{SH} & & \text{NH}_2 \\
| & & | \\
\text{CH}_2 & \quad\text{CH} & \text{CH}_2 \\
| & +\ \text{HN}\diagdown\|\quad\longrightarrow & | \\
-\text{NH}-\text{CH}-\text{CO}- & \quad\text{CH} & \text{CH}_2 \\
& & | \\
& & \text{S} \\
& & | \\
& & \text{CH}_2 \\
& & | \\
& & -\text{NH}-\text{CH}-\text{CO}-
\end{array}
$$

which, being similar to the lysine side chain fits into the active site of trypsin. Accordingly cleavage of the chain will occur at the modified cysteine residue.

Next to trypsin *chymotrypsin* is the most preferred proteolytic enzyme in sequencing. Its specificity is less absolute than that of trypsin. Primarily the bonds that follow phenylalanine, tyrosine and tryptophan are cleaved, but measurable hydrolysis takes place next to leucine and methionine residues as well. It is advisable, therefore, to determine in preliminary experiments the conditions (enzyme-substrate ratio, time, temperature) best suited for the formation of a few and well separable fragments. Occasionally also less specific enzymes, such as pepsin, papain or thermolysin find application in structure elucidation. For the hydrolysis of specific bonds new microbial proteases can be isolated. There are known prolidases and also enzymes which hydrolyze solely the bond which follows a pyroglutamyl residue and so on.

Determination of the amino acid sequence of the gastrointestinal hormone secretin (Mutt et al. 1970) is a good example of the application of selective cleavage of peptide chains with the help of proteolytic enzymes. Amino acid analysis revealed a chain of 27 residues among them four arginines. No lysine was present in the hydrolysate. End group analysis showed that the single histidine residue is in the N-terminal position. No C-terminal residue with a free carboxyl group was found. Tryptic digestion yielded the expected 5 fragments. The largest of them, a 12-peptide contained both histidine and arginine. Thus it must have originated from the N-terminal part of the secretin chain and from the way it was obtained followed that arginine is C-terminal in the fragment. Since it contained phenylalanine, chymotrypsin was used for further fragmentation and the two smaller peptides formed were subjected to Edman degradation. These experiments established the sequence of the first 12 amino acid residues in secretin as His-Ser-Asp-Gly-Thr-Phe-Thr-Ser-Glu-Leu-Ser-Arg. A second tryptic fragment contained no arginine, hence it stemmed from the C-terminal end of the parent molecule. Its sequence, determined by Edman degradation, Leu-Leu-Gln-Gly-Leu-Val-NH$_2$, with valinamide at the C-terminus, explained why no C-terminal residue could be found by end-group analysis in secretin. From the remaining 3 tryptic fragments the smallest, a dipeptide, required only amino acid analysis for determination of its short sequence. Since it was obtained through trypsin catalyzed hydrolysis arginine had to be the C-terminal residue and therefore it could not have any other sequence than

Leu-Arg. A tripeptide fragment was found to be Leu-GlX-Arg via amino acid analysis and determination of the N-terminal residue. The letter "X" in GlX denotes the uncertainity in the results: amino acid analysis of acid hydrolysates (even if a mole of ammonia is found) fails to distinguish firmly between glutamic acid and glutamine. Degradation with leucineaminopeptidase, however, leaves side chain carboxamides intact and could thus be used in the final assignment of Leu-Gln-Arg. The last tryptic fragment, Asp-Ser-Ala-Arg was sequenced by Edman degradation. The remaining major problem, determination of the position of the dipeptide, tripeptide and tetrapeptide fragments in the chain of the parent molecule, was solved through the sophisticated use of selective cleavage with proteolytic enzymes. When digestion with trypsin was interrupted after ten minutes, hydrolysis of one of the bonds that follow arginine residues was far from complete. The Arg-Asp linkage is relatively resistant to trypsin and, therefore, the hexapeptide Leu-Arg-Asp-Ser-Ala-Arg could be isolated from the digest. Yet, this information was still insufficient for determination of the sequence of the entire chain and additional fragmentation was needed to provide the missing evidence for the order of tryptic fragments. Hydrolysis catalyzed by thrombin, a proteolytic enzyme selectively cleaved a single bond in secretin, the same bond that was relatively resistant to trypsin. A brief study of the thrombinic fragments (end group analysis and amino acid analyses) established the complete amino acid sequence of porcine secretin:

His – Ser – Asp – Gly – Thr – Phe – Thr – Ser – Glu – Leu – Ser – Arg $\downarrow$ Leu – Arg $\downarrow$
 1        2        3        4        5        6        7        8        9        10       11       12              13       14

Asp – Ser – Ala – Arg $\downarrow$ Leu – Gln – Arg $\downarrow$ Leu – Leu – Gln – Gly – Leu – Val – NH$_2$
 15       16       17       18              19       20       21              22       23       24       25       26       27       28

The bonds cleaved by trypsin are indicated with arrows. The bond between residues 14 and 15 is the one relatively resistant to trypsin but hydrolyzed by thrombin. Correctness of the structure elucidated by degradation could be corroborated by synthesis (Bodanszky et al. 1967). The synthetic material was indistinguishable from the natural hormone in a series of experiments including the determination of biological activity. Side by side degradation of natural and synthetic samples with trypsin and by chymotrypsin followed by comparison of the fragments on chromatograms and electropherograms completed the study.

## 2. Chemical Methods of Cleavage

Partial acid hydrolysis is generally non-specific but can still be exploited for the cleavage of the peptide bond at aspartyl residues. The Asp-Pro linkage is exceptionally sensitive to acids and is cleaved by dilute hydrochloric acid at room temperature. At elevated temperatures very weak acids such as 0.03 M

HCl or 0.25 M acetic acid are sufficient to effect hydrolysis of the peptide bonds on both sides of aspartyl residues. For instance the hexapeptide Val-Leu-Gly-Asp-Phe-Pro yields the tripeptide Val-Leu-Gly, the dipeptide Phe-Pro and free aspartic acid. Release of aspartic acid is usually complete after a day of hydrolysis at 100 °C in 0.25 M AcOH while hydrolytic fission of other peptide bonds is negligible. Asparagine creates a notable exception because the carboxamide group in its side chain is fairly sensitive to acid catalyzed hydrolysis and the free carboxyl group of the gradually formed aspartyl residue provides the neighboring group effect which leads to the excision of aspartic acid from the chain.

A more frequently applied method of selective cleavage is based on the reaction of tyrosine and tryptophan side chains with positively charged bromine. The first observations involved treatment of tyrosine containing peptides with bromine water. Hypobromous acid in the latter generates a tribromoderivative of the phenolic side chain in tyrosine which then rearranges to a spirolactone with the concomitant cleavage of the peptide bond:

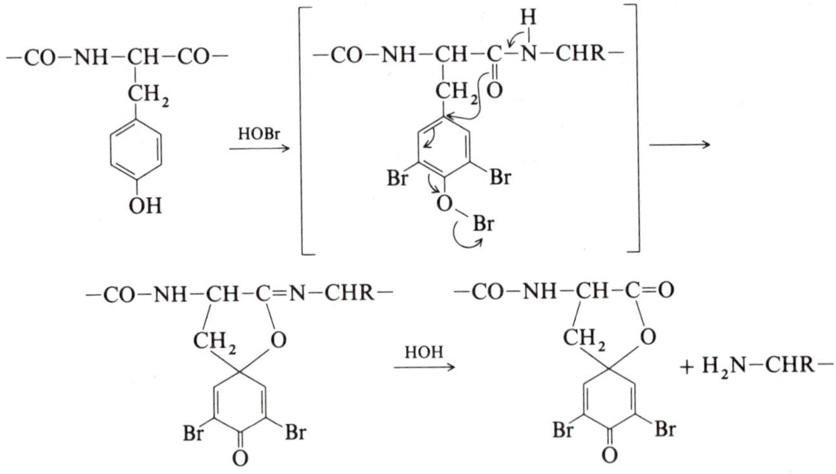

The reaction is readily monitored by the characteristic changes in the uv spectrum. A more reliable source of positively charged bromine is N-bromosuccinimide, which has often been used for cleavage at tryptophan:

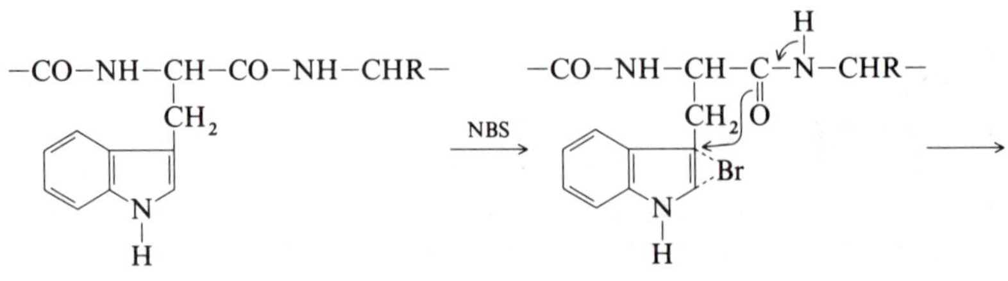

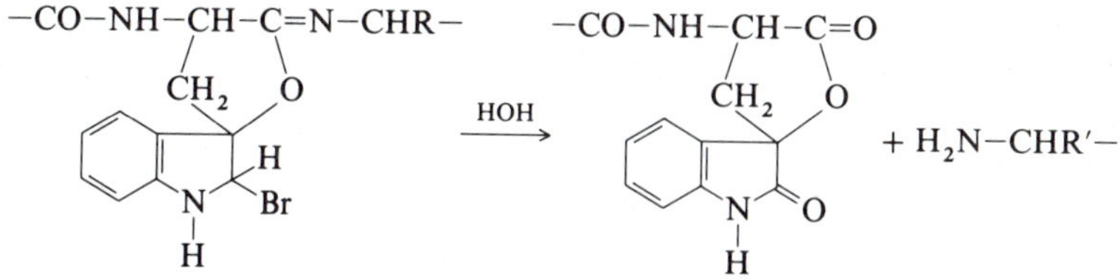

$$-CO-NH-CH-C=N-CHR-$$

with CH$_2$, O side chain and indoline-Br structure

→ (HOH) →

$$-CO-NH-CH-C=O$$

with CH$_2$, O side chain and oxindole structure  $+ H_2N-CHR'-$

Of course the tyrosine side-chain is not inert toward N-bromosuccinimide and to some extent also histidine residues are affected. A more selective reagent for the cleavage at tryptophan was sought and found in the skatol derivative

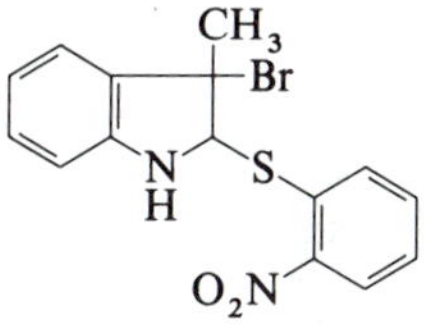

(indoline structure with CH$_3$, Br, N-H, S, and $O_2N$ nitrophenyl group)

as the source for positively charged bromine. Oxidative fission e.g. with o-iodo-sobenzoic acid (Fontana et al. 1983) has also been proposed.

A somewhat analogous fission of the peptide chain occurs when methionine containing peptides are exposed to the action of cyanogen bromide:

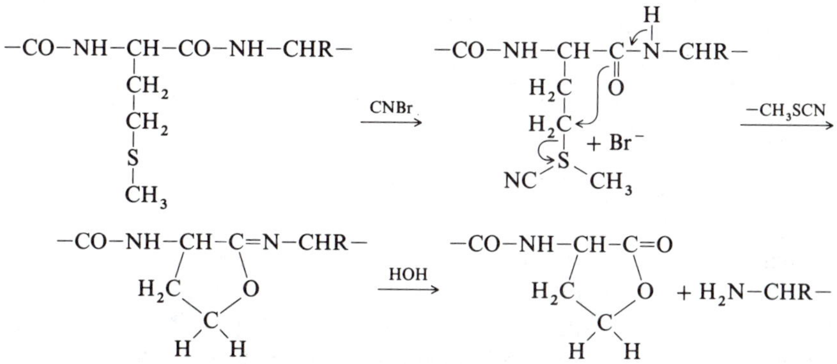

$$-CO-NH-CH-CO-NH-CHR-$$

with CH$_2$, CH$_2$, S, CH$_3$ side chain

→ (CNBr) →

$$-CO-NH-CH-C\overset{H}{\underset{}{N}}-CHR-$$

with H$_2$C, H$_2$C, C, S, CH$_3$, NC side chain and + Br$^-$

→ ($-CH_3SCN$) →

$$-CO-NH-CH-C=N-CHR-$$

with H$_2$C, O, C, H, H side chain

→ (HOH) →

$$-CO-NH-CH-C=O$$

with H$_2$C, O, C, H, H side chain  $+ H_2N-CHR-$

Selectivity in the cyanogenbromide induced fission (Gross and Witkop 1962) is quite exceptional and no major side-reactions accompany the process. It is understandable, therefore, that once a methionine residue is found by analysis the method is routinely applied. One concern must be voiced however. The homoserine lactone formed in the reaction is not inert toward nucleophiles and will act as an acylating agent if amines are present. A reestablishment of the peptide bond cleaved by cyanogen bromide is mostly negligible, but if the two newly generated chain-fragments are held together by a disulfide bond or by a combination of forces such as ion-pair formation, hydrogen bonds or non-polar interaction, then an intramolecular or quasi-intramolecular reaction will occur, producing a chain in which the methionine residue of the parent molecule is replaced by homoserine:

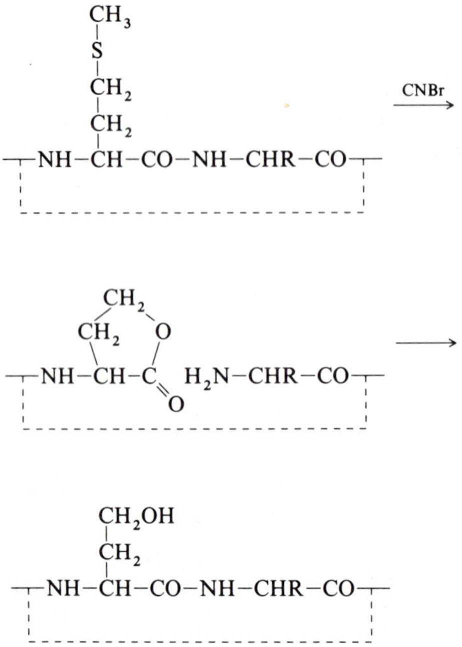

$$\underset{\text{CNBr}}{\longrightarrow}$$

$$\begin{array}{c} CH_3 \\ | \\ S \\ | \\ CH_2 \\ | \\ CH_2 \\ | \\ -NH-CH-CO-NH-CHR-CO- \end{array}$$

$$\longrightarrow$$

$$-NH-CH-C\underset{O}{\overset{CH_2}{\underset{CH_2}{\diagup}}}\,O \quad H_2N-CHR-CO-$$

$$\begin{array}{c} CH_2OH \\ | \\ CH_2 \\ | \\ -NH-CH-CO-NH-CHR-CO- \end{array}$$

## 3. Fission of Disulfides

Cysteine containing peptides require a special operation before sequencing. Even if originally free sulfhydryl groups were present, exposure to air converts them to disulfides which then complicate interpretation of the results of sequence determination. If a single cysteine residue occurs in a molecule, as for instance in the oxydized form of glutathione, then two identical chains are connected through a disulfide bridge:

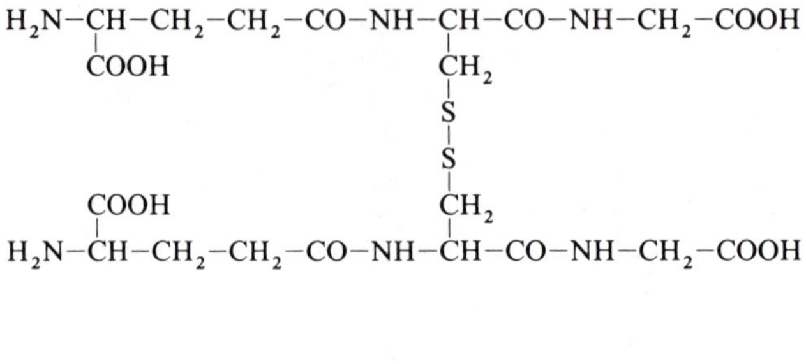

$$H_2N-CH-CH_2-CH_2-CO-NH-CH-CO-NH-CH_2-COOH$$
$$\qquad\quad | \qquad\qquad\qquad\qquad\qquad | $$
$$\qquad\quad COOH \qquad\qquad\qquad\qquad CH_2$$
$$\qquad\qquad\qquad\qquad\qquad\qquad\qquad\quad | $$
$$\qquad\qquad\qquad\qquad\qquad\qquad\qquad\quad S$$
$$\qquad\qquad\qquad\qquad\qquad\qquad\qquad\quad | $$
$$\qquad\qquad\qquad\qquad\qquad\qquad\qquad\quad S$$
$$\qquad\quad COOH \qquad\qquad\qquad\qquad CH_2$$
$$\qquad\quad | \qquad\qquad\qquad\qquad\qquad | $$
$$H_2N-CH-CH_2-CH_2-CO-NH-CH-CO-NH-CH_2-COOH$$

In peptides containing two cysteines a cyclic disulfide, such as the one found in oxytocin

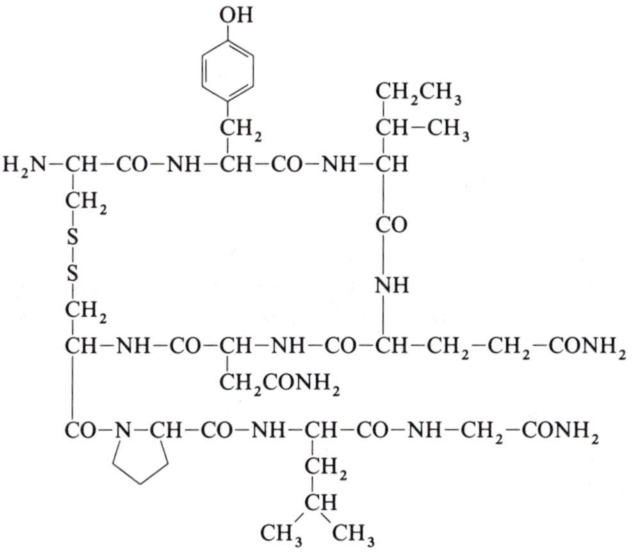

will form by intramolecular reaction. Oxidation of the chain in solutions of higher concentration, however, will produce through intermolecular reaction two dimers as well, a parallel and an antiparallel dimer

$$H-Cys-Tyr-Ile-Gln-Asn-Cys-Pro-Leu-Gly-NH_2$$
$$|\qquad\qquad\qquad\qquad\qquad |$$
$$H-Cys-Tyr-Ile-Gln-Asn-Cys-Pro-Leu-Gly-NH_2$$

$$H-Cys-Tyr-Ile-Gln-Asn-Cys-Pro-Leu-Gly-NH_2$$
$$|\qquad\qquad\qquad\qquad\qquad |$$
$$H_2N-Gly-Leu-Pro-Cys-Asn-Gln-Ile-Tyr-Cys-H$$

(H – Cys – and Cys – H – indicate N-terminal cysteine residues)

and also oligomers in which three or more chains are linked together through disulfide bridges.

Both intrachain and interchain disulfide bridges are present in insulin (cf. Table 3, p. 7) and it is obvious that determination of the sequence of the two chains had to be preceded by their separation.

There is an almost embarrassingly rich choice of methods by which disulfides can be cleaved, for instance in a smooth reaction with potassium cyanide or with sodium hydrogen sulfite (bisulfite)

$$
\begin{array}{ccc}
\begin{array}{l}
-\text{NH}-\text{CH}-\text{CO}- \\
\quad\quad\ |\\
\quad\quad\ \text{CH}_2 \\
\quad\quad\ |\\
\quad\quad\ \text{SH} \\
\quad\quad\ |\\
\quad\quad\ \text{S}-\text{SO}_3\text{Na} \\
\quad\quad\ |\\
\quad\quad\ \text{CH}_2 \\
-\text{NH}-\text{CH}-\text{CO}-
\end{array}
&
\begin{array}{l}
-\text{NH}-\text{CH}-\text{CO}- \\
\quad\quad\ |\\
\quad\quad\ \text{CH}_2 \\
\quad\quad\ |\\
\quad\quad\ \text{S} \\
\quad\quad\ |\\
\quad\quad\ \text{S} \\
\quad\quad\ |\\
\quad\quad\ \text{CH}_2 \\
-\text{NH}-\text{CH}-\text{CO}-
\end{array}
&
\begin{array}{l}
-\text{NH}-\text{CH}-\text{CO}- \\
\quad\quad\ |\\
\quad\quad\ \text{CH}_2 \\
\quad\quad\ |\\
\quad\quad\ \text{SCN} \\
\quad\quad\ |\\
\quad\quad\ \text{SK} \\
\quad\quad\ |\\
\quad\quad\ \text{CH}_2 \\
-\text{NH}-\text{CH}-\text{CO}-
\end{array}
\end{array}
$$

(left arrow: NaHSO$_3$; right arrow: KCN)

but in the praxis of structure determination only oxidation and reduction are applied. Treatment of formic acid with hydrogen peroxide in the cold generates performic acid which cleaves the cystine moiety in peptides with the formation of two cysteic acid residues:

$$
\begin{array}{cc}
\begin{array}{l}
-\text{NH}-\text{CH}-\text{CO}- \\
\quad\quad\ |\\
\quad\quad\ \text{CH}_2 \\
\quad\quad\ |\\
\quad\quad\ \text{S} \\
\quad\quad\ |\\
\quad\quad\ \text{S} \\
\quad\quad\ |\\
\quad\quad\ \text{CH}_2 \\
-\text{NH}-\text{CH}-\text{CO}-
\end{array}
&
\begin{array}{l}
-\text{NH}-\text{CH}-\text{CO}- \\
\quad\quad\ |\\
\quad\quad\ \text{CH}_2 \\
\quad\quad\ |\\
\quad\quad\ \text{SO}_3\text{H} \\
\quad\quad\ |\\
\quad\quad\ \text{SO}_3\text{H} \\
\quad\quad\ |\\
\quad\quad\ \text{CH}_2 \\
-\text{NH}-\text{CH}-\text{CO}-
\end{array}
\end{array}
$$

(arrow: HCO$_3$H)

A serius drawback of this otherwise simple and straightforward method of chain separation is the damage suffered by tyrosine and tryptophan side chains in the process. Oxidation of methionine residues to the corresponding sulfoxides also occurs, but it is reversible. In the presence of heavy metal impurities, however, irreversible oxidation to sulfones can take place

$$
\begin{array}{ccc}
\begin{array}{l}
-\text{NH}-\text{CH}-\text{CO}- \\
\quad\quad\ |\\
\quad\quad\ \text{CH}_2 \\
\quad\quad\ |\\
\quad\quad\ \text{CH}_2 \\
\quad\quad\ |\\
\quad\quad\ \text{O}=\text{S}=\text{O} \\
\quad\quad\ |\\
\quad\quad\ \text{CH}_3
\end{array}
&
\begin{array}{l}
-\text{NH}-\text{CH}-\text{CO}- \\
\quad\quad\ |\\
\quad\quad\ \text{CH}_2 \\
\quad\quad\ |\\
\quad\quad\ \text{CH}_2 \\
\quad\quad\ |\\
\quad\quad\ \text{S} \\
\quad\quad\ |\\
\quad\quad\ \text{CH}_3
\end{array}
&
\begin{array}{l}
-\text{NH}-\text{CH}-\text{CO}- \\
\quad\quad\ |\\
\quad\quad\ \text{CH}_2 \\
\quad\quad\ |\\
\quad\quad\ \text{CH}_2 \\
\quad\quad\ |\\
\quad\quad\ \text{S}=\text{O} \\
\quad\quad\ |\\
\quad\quad\ \text{CH}_3
\end{array}
\end{array}
$$

(left arrow: oxid. cat; middle arrows: oxid. / red.)

These complicating factors do not render the performic acid method impractical, but they must be taken into account in the interpretation of the results of subsequent degradations.

Numerous reducing agents have been applied for the cleavage of disulfides in peptides. Reduction with sodium in liquid ammonia, with cysteine, 2-mercaptoethanol, 2-mercaptoacetic acid, 2-mercaptoethylamine (cysteamine), borohydrides and trialkylphosphines were used both for preparative and for analytical purposes. In degradation studies the two reagents of Cleland (1964), dithiothreitol and dithioerythritol

$$\begin{array}{cc}
\begin{array}{l}
CH_2-SH \\
\;\,| \\
H-C-OH \\
\;\,| \\
HO-C-H \\
\;\,| \\
CH_2-SH
\end{array}
&
\begin{array}{l}
CH_2-SH \\
\;\,| \\
HO-C-H \\
\;\,| \\
HO-C-H \\
\;\,| \\
CH_2-SH
\end{array}
\end{array}$$

are quite convenient. Their usefulness stems from the shift in equilibrium due to the stability of cyclic disulfides formed from the reagents

$$\begin{array}{l}
-NH-CH-CO- \\
\quad\quad\;| \\
\quad\quad\;CH_2 \\
\quad\quad\;| \\
\quad\quad\;S \\
\quad\quad\;| \\
\quad\quad\;S \\
\quad\quad\;| \\
\quad\quad\;CH_2 \\
\quad\quad\;| \\
-NH-CH-CO-
\end{array}
\; + \;
\begin{array}{l}
CH_2-SH \\
\;\,| \\
H-C-OH \\
\;\,| \\
HO-C-H \\
\;\,| \\
CH_2-SH
\end{array}
\;\rightleftharpoons\;
\begin{array}{l}
-NH-CH-CO- \\
\quad\quad\;| \\
\quad\quad\;CH_2 \\
\quad\quad\;| \\
\quad\quad\;SH \\
\quad\quad\;| \\
\quad\quad\;SH \\
\quad\quad\;| \\
\quad\quad\;CH_2 \\
\quad\quad\;| \\
-NH-CH-CO-
\end{array}
\; + \; H_2C
\begin{array}{c}
OH\;\;H \\
\backslash C-C \diagup \\
\diagup H\;\;OH \backslash CH_2 \\
\backslash S=S \diagup
\end{array}$$

Since exposure to air reconverts the sulfhydryl groups to disulfides, it is necessary to stabilize the reduced forms by alkylation with benzyl chloride or with iodoacetic acid

$$\begin{array}{l}
-NH-CH-CO- \\
\quad\quad\;| \\
\quad\quad\;CH_2 \\
\quad\quad\;| \\
\quad\quad\;S \\
\quad\quad\;| \\
\quad\quad\;CH_2-COOH
\end{array}
\;\xleftarrow{\;I-CH_2COOH\;}\;
\begin{array}{l}
-NH-CH-CO- \\
\quad\quad\;| \\
\quad\quad\;CH_2 \\
\quad\quad\;| \\
\quad\quad\;SH
\end{array}
\;\xrightarrow{\bigcirc-CH_2Cl}\;
\begin{array}{l}
-NH-CH-CO- \\
\quad\quad\;| \\
\quad\quad\;CH_2 \\
\quad\quad\;| \\
\quad\quad\;S \\
\quad\quad\;| \\
\quad\quad\;CH_2-\bigcirc
\end{array}$$

Finally, complete structure determination also requires that the positions of disulfide bridges — if there is more than one — be established. Partial hydrolysis can provide fragments with an intact disulfide bridge. In the process the possibility of disulfide dismutation

$$2\,R-S-S-R' \;\rightleftharpoons\; R-S-S-R + R'-S-S-R'$$

which is catalyzed by traces of mercaptanes and also by cyanide and by bicarbonate anions, should not be overlooked, particularly because the disulfide interchange reaction has a tendency for the generation of symmetrical disulfides.

# References

Akabori, S., Ohno, K., Narita, K.: Bull. Chem. Soc. Japan *25*, 214 (1952)

Bergmann, M., Miekeley, A., Kann, E.: Liebigs Ann. Chem. *458*, 40 (1927)

Bodanszky, M., Ondetti, M. A., Levine, S. D., Williams, N. J.: J. Amer. Chem. Soc. *89*, 6753 (1967)

Cleland, W. W.: Biochemistry *3*, 480 (1964)

Dakin, H. D., West, R.: J. Biol. Chem. *78*, 91, 745, 757 (1928)

Edman, P.: Acta Chem. Scand. *4*, 283 (1950)

Edman, P., Begg, G.: Eur. J. Biochem. *1*, 80 (1967)

Fromageot, C., Jutisz, M., Mayer, D., Penasse, L.: Biochem. Biophys. Acta *6* 283 (1950)

Fontana, A., Dalzoppo, D., Grandi, C., Zamboni, M. in Methods in Enzymology, Vol. 91 (Hirs, C. W. H. and Timasheff, S. N. eds.) Academic Press, New York 1983, p. 311

Gray, W. R., Hartley, B. S.: Biochem. J. *89*, 59P, 379 (1963)

Gross, E., Witkop, B.: J. Biol. Chem. *237*, 1856 (1962)

Mutt, V., Jorpes, J. E., Magnusson, S.: Eur. J. Biochem. *15*, 513 (1970)

Sanger, F.: Biochem. J. *39*, 507 (1945)

# Additional Sources

Greenstein, J. P. and Winitz, M., Sequential Analysis of Peptides in Chemistry of the Amino Acids, p. 1512. New York, J. Wiley and Sons 1961

Gray, W. R., Dansyl Chloride Procedure in Methods of Enzymology, Vol. 11 (Hirs, C. H. E., ed) pp. 139–151, New York, Acad. Press 1967

Ambler, R. A.: Enzymic Hydrolysis with Carboxypeptidase, ibid. pp. 155–156

# IV. Secondary and Tertiary Structure

## A. Architectural Features in Peptides

### 1. The Peptide Bond

The bond between the carbon atom of the carbonyl group and the amide nitrogen has partial double bond character that can be attributed to resonance:

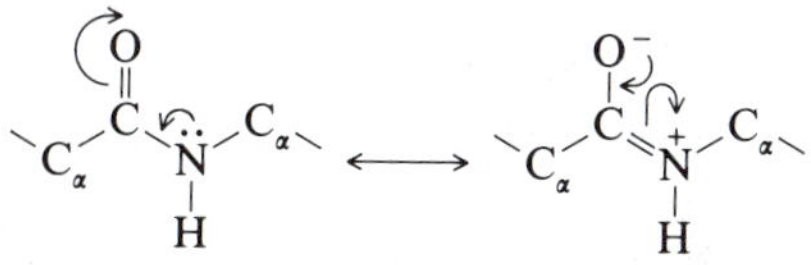

Accordingly, the carbonyl carbon and oxygen atoms, the amide nitrogen and hydrogen and the two adjacent α-carbon atoms lie approximately in the same plane. Experimental evidence, mainly from x-ray crystallography supports the near coplanarity of these six atoms in peptide backbones and also shows a relatively short distance between the carbonyl carbon and the nitrogen. The carbonyl oxygen and the amide nitrogen are on opposite sides of the (partial) double bond, at least in most peptide bonds. The cis arrangement is somewhat less stable and since there is a considerable energy barrier between the cis and trans forms, the latter generally prevails.

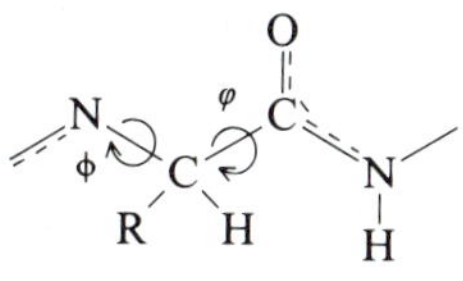

The partial double bond character of each third bond in the peptide backbone would cause some limitation in the possible geometries of the chain but this in itself is certainly not sufficient to determine a preferred conformation. A well defined architecture requires further limiting factors and such can be found in the restricted rotation around the $N-C_\alpha$ and the $C_\alpha-C$ bonds. Negative (repulsive) interaction between side chains of neighboring amino acid residues hinders free rotation around these bonds and the dihedral angles $\Phi$ and $\Psi$, corresponding to rotation out of the plane around the $N-C_\alpha$ and $C_\alpha-C$ bonds respectively can take up only certain values. The allowable angles were calculated (Ramachandran et al. 1963) and are represented in "Ramachandran plots". These calculations and plots were vindicated in numerous studies of peptides and proteins by x-ray crystallography.

## 2. Secondary Structures

Polarization of the carbonyl group results in a partial positive charge at the oxygen atom, because it is more electronegative than the carbon atom. In the N–H bond of the amide group the nitrogen is the more electro-negative atom, hence a partial positive charge rests on the hydrogen. If the CO and NH groups are in sufficient proximity, a positive (attractive) interaction results, a "hydrogen bond" in which one of the unshared pairs of electrons of the oxygen atom provides a weak bonding effect

$$\overset{\backslash}{\underset{/}{C}} \overset{\delta+}{=} \overset{\delta-}{\underset{\cdot\cdot}{O}} : \quad \overset{\delta+}{H} \overset{\delta-}{-} \overset{/}{\underset{\backslash}{N}} \qquad\qquad \overset{\backslash}{\underset{/}{C}} = O \cdots\cdot H - \overset{/}{\underset{\backslash}{N}}$$

amounting to a few kilocalories per mole. In the hydrogen bond the hydrogen atom belongs, to some extent, to both electronegative atoms. In addition to oxygen and nitrogen, also atoms of sulfur and fluorine can serve as bridgehead. Hydrogen bonds are of extreme importance in the three dimensional structure of peptides and proteins. The cumulative effect of many weak attractive forces along the peptide backbone create well defined geometries and the forms stabilized by hydrogen bonds are called "secondary structure".

   Early studies by Pauling, Corey and Branson (1951), involving x-ray structures of small peptides and experimentation with molecular models, resulted in the prediction of several geometric patterns in proteins. Of these the *α-helix* and *β-sheets* were subsequently encountered as dominant architectural features in numerous proteins. In more recent times also *reverse turns* (hairpin turns, *γ*-turns) were postulated (Venkatachalam 1968) and then recognized as frequent contributors to peptide and protein conformation.

   The *α-helix* is a spiral structure stabilized by intramolecular hydrogen bonds between carbonyl oxygens and amide nitrogens four residues apart:

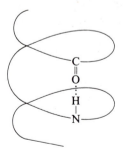

The hydrogen bonds are parallel to the axis of the helix and close a ring of 14 atoms including the hydrogen. There are thus 3.6 residues in one turn of the spiral which has a "pitch" (distance between turns) of 5.4 A. Accordingly a 5.4:3.6 = 1.5 A spacing was observed in x-ray diffraction patterns of powders of certain helical polyamino-acids and also of proteins of high helix content. This means that the same groups (CO, NH) repeatedly occur with a distance of 1.5 A between identical groups. It should be noted however, that there are several other, less frequently encountered helices, with different dimensions. For instance an elongated helix is present in polyproline and in the proline-rich protein collagen.

Peptide chains (or two parts of the same chain) connected with each other through multiple hydrogen bonds form *β-sheets*. The participating chains can be parallel or antiparallel to each other. An antiparallel sheet is shown here including a view along the general plane of the sheet to indicate its pleated character:

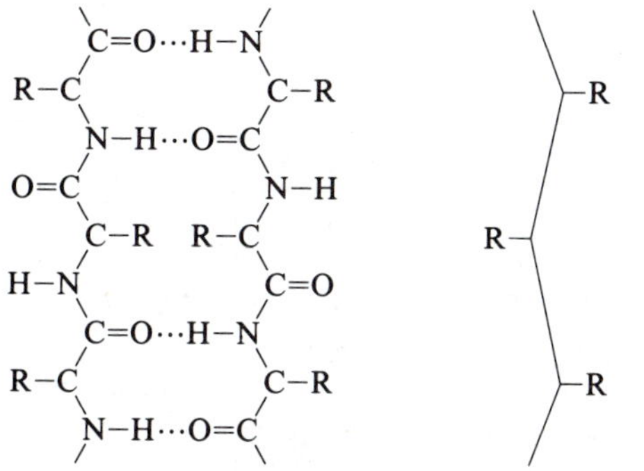

$$
\begin{array}{cc}
\text{C=O}\cdots\text{H--N} \\
\text{R--C} \qquad \text{C--R} \\
\text{N--H}\cdots\text{O=C} \\
\text{O=C} \qquad \text{N--H} \\
\text{C--R} \quad \text{R--C} \\
\text{H--N} \qquad \text{C=O} \\
\text{C=O}\cdots\text{H--N} \\
\text{R--C} \qquad \text{C--R} \\
\text{N--H}\cdots\text{O=C}
\end{array}
\qquad
\begin{array}{c}
\text{--R} \\
\text{R--} \\
\text{--R}
\end{array}
$$

In *reverse turns* a sharp change in the direction of the peptide backbone is stabilized by an intramolecular hydrogen bond between a carbonyl oxygen atom and a nearby amide nitrogen. The ring thus formed can contain 5 or 8 or 11 etc. atoms and accordingly the conformation is designed as $C_5$ or $C_8$ or $C_{11}$ etc.:

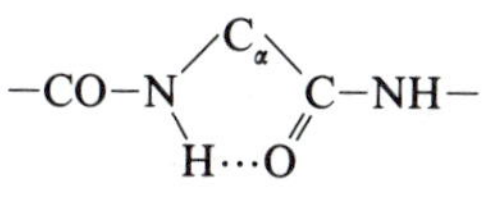

$$
\begin{array}{cc}
\text{--CO--N} \overset{\displaystyle C_\alpha}{\diagup\diagdown} \text{C--NH--} \\
\text{H}\cdots\text{O} \\
C_5
\end{array}
\qquad
\begin{array}{c}
\text{--CO--N--C}_\alpha\text{--CO} \\
\text{H} \\
\vdots \\
\text{O} \\
\text{--NH--C--C}_\alpha\text{--NH} \\
C_8
\end{array}
$$

$$
\begin{array}{c}
\text{--CO--N--C}_\alpha\text{--CO--NH} \\
\text{H} \qquad\qquad\quad \text{C}_\alpha \\
\vdots \\
\text{O} \\
\text{--NH--C--C}_\alpha\text{--NH--CO} \\
C_{11}
\end{array}
$$

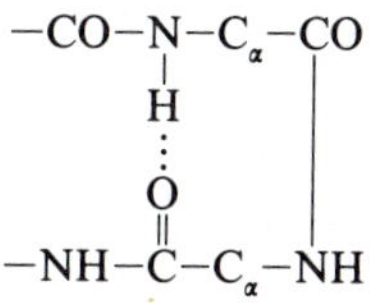

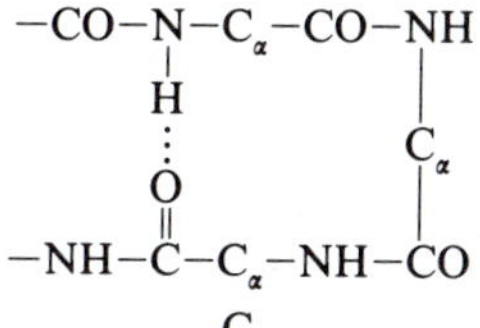

If one of the amino acid residues in a reverse turn has a configuration different from that of the other residues, then the hydrogen bond formed is more stable. In proteins and in the majority of biologically active peptides (which are usually cleavage products of proteins) only L-residues are present, but in microbial peptides D-residues are quite common. It is not surprising, therefore, that most of them have cyclic structures: ring closure is very much facilitated by reverse turns.

Stabilization of conformation by multiple hydrogen bonds is a *cooperative effect*. Disruption of a single hydrogen bond is not enough for destabilization, but the secondary structure collapses if a major part of the hydrogen bonds is disrupted. If compounds such as urea or guanidine salts, that are able to compete for the intramolecular hydrogen bonds are added to aqueous solutions of peptides or proteins the secondary structure disappears. The change does not take place gradually: at a certain urea etc. concentration a sudden collapse is noted: the structure "melts".

## 3. Tertiary Structure

Well defined folding of a peptide chain is called its tertiary structure. Hydrogen bond-stabilized reverse turns can contribute to folding but they are not its primary cause. A more important factor in folding was recognized in *non-polar interaction*, often described as *"hydrophobic bond"* (Kauzmann 1959).

When bulky side chains of aliphatic or aromatic amino acids approach each other a negative interaction (repulsion) would be expected between them. In aqueous solution, however, a positive interaction (attraction) is observed that can be explained with the escape of water molecules from the cage formed by non-polar regions. Inside such a cage water molecules are linked to each other by hydrogen bonds but obviously not to the surrounding non-polar amino acid side chains. These water molecules are, therefore, in a highly ordered state and order will decrease if they escape from the cage to join other water molecules outside. As the net effect of decrease in order (increase in entropy) the non-polar side chains are held close to each other by a "hydrophobic bond".

The importance of non-polar interaction can not be overemphasized. Hydrogen bonds which stabilize secondary structures should not exist in aqueous solutions of peptides and proteins. In the presence of a large number of water molecules the equilibrium

$$\diagdown\!\!C=O\cdots H-N\diagup + nH_2O \ \rightleftarrows \ \diagdown\!\!C=O\cdots H \diagdown\!\!O + \quad \overset{H-N}{\underset{H}{\vdots}} \overset{}{\underset{O}{}} \diagdown H$$

is expected to shift to the right and the hydrogen bonds between carbonyl carbons and amide nitrogens to be eliminated. Yet, non-polar interaction generates "hydrophobic pockets" in which there is no competition by water molecules for hydrogen bonds.

On one side of helices typically non-polar residues are found, with their side chains in contact with other hydrophobic residues further along the peptide chain. The attractive forces between such non-polar groups lead to a general folding of the chain, its "tertiary structure". The hydrophobic region created by the non-polar contacts within the fold contribute to the stability of the helix. Hence the expressions "secondary structure" and "tertiary structure" could be replaced by the term "secondary-tertiary structure". This interdependence between the two kind of structures extends to other hydrogen bond stabilized geometries, to β-sheets and reverse turns as well. Yet, stabilization of secondary structure by non-polar interaction can take place also without folding. Thus, hydrophobic bonds are found not solely between two sections of the same chain but also between two separate chains. Furthermore, intermolecular stabilization can be effected by molecules other than peptides. In numerous peptides helices are generated on addition of water-miscible alcohols, such as methanol, ethanol, isopropanol, tert.butanol, chloroethanol and particularly trifluoroethanol or hexafluoroisopropanol, to their aqueous solutions. Similar effects were observed on addition of detergents. These phenomena can be rationalized by the displacement of water from the proximity of amide groups and the ensuing elimination of competition for hydrogen bonds.

The role of disulfides in the determination of tertiary structure is less important than it would a priori appear. The architecture of peptides and proteins is primarily determined by the interaction of side chains of residues which are next neighbors and by long range cooperative effects of non-polar interactions. Disulfides form merely between sulfhydryl groups that are already in each others proximity. This was impressively demonstrated in the reduction and reoxidation of ribonuclease (White 1961). Instead of the formation of numerous disulfides different from those present in the parent molecule, the original disulfides were restored and the enzyme was regenerated with full biological activity. Also, random oxidation of six sulfhydryl groups in the two separated chains of insulin should produce a vast number of different molecules, but if prior to oxidation the two chains are first allowed to interact in aqueous solution then mainly the three disulfide bridges characteristic for insulin are formed

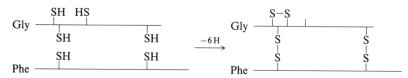

and insulin is the principal product.

A similar limited role can be assigned to *polar (coulombic) interaction*. Ion pair formation between basic and acidic side chains is quite commonly observed in peptides and proteins. These "salt bridges", however, are not primary determinants of architecture. For instance, in secretin (conf. page 7) two ion pairs can be discerned, but in synthetic analogs with only one or no ion pair the original geometry of the molecule is not grossly altered. Thus, the contribution of disulfide bridges and ion pairs to the final determination of conformation in peptides consists mainly of providing certain rigidity to an already existing architecture.

## 4. Quaternary Structure

The expression "quaternary structure" is usually applied only for proteins. Their large molecules, containing hundred or more covalently bound amino acid residues, can assemble into still larger structures in which two or several "subunits" are linked through hydrogen bonds, polar or non-polar interactions. This kind of aggregation is known in myoglobin, hemoglobin and in numerous enzymes, all globular proteins. It appears, that at least some peptides are similarly prone to self-association. Thus insulin crystallizes in the form of a hexamer of the covalently built molecule while a trimer was revealed by x-ray crystallography in glucagon. (The primary structure of insulin and that of glucagon are shown on p. 7.) In both hormones self-association was accompanied by major conformational changes. In dilute aqueous solutions neither insulin nor glucagon exhibit pronounced helical character but molecules with high helix content are present in the crystals. The newly formed helical structures are the consequence of hydrophobic contacts between subunits and the ensuing displacement of water. Glucagon has a general tendency for self-association: under certain conditions $\beta$-structures appear in solution and aggregation can proceed to the point where insoluble fibrils separate. Fibril formation is known for insulin as well.

Self association seems to be related to a degree of conformational freedom. Smaller peptides with well defined architecture, such as oxytocin have not been observed in aggregated form so far. On the other hand many blocked intermediates in peptide synthesis exhibit annoying insolubility in organic solvents because of their tendency for aggregation in the form of $\beta$-sheets.

# B. Methods for the Analysis of Conformation of Peptides

Electronic spectra yield valuable information on the presence or absence of chromophores and functional groups, but have rather limited use in the elucidation of the three dimensional structure in peptides. Infrared spectroscopy has been applied for the detection of helices and $\beta$-sheets, yet the spectra are usually meaningful only when the molecules are somewhat ordered as, for instance, in stretched films of polyamino acids. The scope of investigations seems to broaden since the advent of laser-Raman spectroscopy. Absorption in the ultraviolet is a similarly modest source of information about peptide architecture. Difference spectra can reveal minor shifts in the absorption maxima of tyrosine in proteins in which the chromophore is perturbed by the proximity of residues far along the peptide chain but brought close by folding. For the smaller molecules of peptides difference spectra were not commonly applied. A more productive approach to conformational analysis of peptides is fluorescence spectroscopy. Among the twenty amino acids which are constituents of proteins only three, phenylalanine, tyrosine and tryptophan are fluorophores (emit fluorescent light), but peptides lacking "intrinsic fluorescence" can be provided with "extrinsic fluorescent" probes by substitution with appropriate groups, for instance through treatment with fluorescamine (cf. 15) or with 5-dimethyl-aminonaphthalene-2-sulfonic acid chloride (cf. page 18). Measurement of fluorescence quantum yields allows conclusions about intramolecular interactions and fluorescence decay studies can reveal side-chain conformational heterogeneity. Most impressive is the determination of intramolecular distances in energy transfer experiments (Förster 1948) in which quenching of the fluorescence of a donor (such as the phenolic group in the tyrosine side chain) and fluorescence enhancement in an acceptor (like the indole system in tryptophan) are determined and from the results the actual physical distance between the two groups calculated. The practical application of fluorescence techniques in the field of peptides was thoroughly described by Schiller (1985).

The probably most extensively used method of conformational analysis in peptide chemistry is the recording of circular dichroism and optical rotatory dispersion spectra. Single crystal x-ray crystallography, when applicable, yields the most detailed information and in recent years nmr spectroscopy emerges as its important competitor. These will be discussed in the following sections. At this point we wish to stress a general shortcoming of all these approaches. They either only give tentative description of the three dimensional structure of the molecule or they describe its geometry in detail, but with validity merely for certain conditions, for instance in solid state or in the presence of detergents. This general limitation is due to a degree of conformational freedom inherent in most peptides. Therefore, it might be overly optimistic to expect that further improvement in the existing methodology can fully overcome this difficulty. A partial solution is the application of several methods of conformational analysis rather than a single one and findings have to be corroborated through the results of additional studies such as viscosimetry or diffusion through membranes and also via the "prediction" of conformation with the help of empirical parameters.

## 1. Optical Rotatory Dispersion and Circular Dichroism

Determination of specific rotation at the sodium D-line (589 nm) is valuable in the characterization of peptides and of intermediates in peptide synthesis. It is routinely applied in the comparison of synthetic peptides with the corresponding natural compounds and also in evaluation of purity but it provides no clue for the three-dimensional structure. The situation becomes entirely different if instead of the determination of optical rotation at a single wavelength, rotation is recorded as the function of wavelength along a broad range of the spectrum, particularly in the ultraviolet. The optical rotatory dispersion (ord) spectrum thus obtained is one of the most valuable tools in the conformational analysis of peptides. The closely related circular dichroism (cd) spectra are based on the difference in absorption between the left and right components of circular polarized light

$$\Delta D = D_l - D_r$$

This difference ($\Delta D$), calculated for a molar solution and one centimeter pathlength is the *molar circular dichroism: $\Delta\varepsilon$*. Quite often instead of $\Delta D$ rather molar ellipticity, $[\Theta]_M$ is shown in graphical representations of cd spectra. This corresponds to the ratio between the short and long axes of the ellipse of the emerging (initially circular) polarized light and is directly proportional to molar circular dichroism

$$[\Theta] = 3300 \, \Delta\varepsilon$$

The two methods give results that are interconvertible hence ord or cd spectra can be selected according to the availability of the appropriate instruments. Yet, one is sometimes preferred over the other when the optical density of the solution in certain areas of the spectrum interferes with the determination of rotation or of dichroism. In both methods the refractive index of the solution should be taken into consideration in the calculation of results, but the ensuing minor correction is often neglected.

Optical activity in peptides stems from two different kinds of chiralities. Asymmetry in the individual amino acid constituents is the lesser contributor while architectural dissymmetry, for instance the handedness of helices, has a major effect on the optical rotatory power of a peptide molecule. This makes it possible to draw conclusions from optical rotation on conformation. Chiroptical (ord and cd) spectra are close to linear in such areas of the spectrum where there is no significant absorption by chromophores and it is possible to calculate the value of rotation at one wavelength from the value determined at another wavelength. The spectra, however, show an anomaly in the neighborhood of absorption bands. In anomalous ord spectra the "Cotton effect" appears: a trough, followed by a peak at a lower wavelength is a negative Cotton effect. The maximum and the minimum lie about 25 nm above and under the absorption maximum of the chromophore which, being in an asymmetric environment, is the cause of the anomaly. In cd spectra a peak appears at the wavelength of the maximum of the chromophore if the Cotton effect is negative and a trough if it is positive. Since the absorption band of the amide

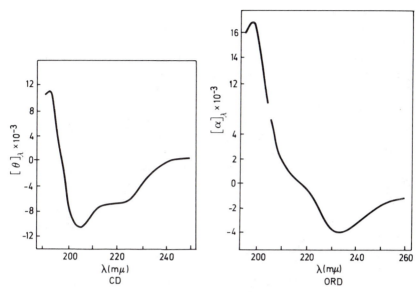

**Fig. 1.** The CD and ORD spectra of secretin in water

group is at about 200 nm, only instruments that allow the recording of spectra even below this wavelength are useful in the study of peptides.

In the ord spectra of polypeptides with established right handed $\alpha$-helical conformation a negative Cotton effect is seen with a trough at 233 nm and a peak at 199 nm.

The value of molar rotation at 233 nm is used as the measure of helix content in peptides, the large peak at 199 nm is less easily recorded. Peptide chains with different types of helices exhibit different ord spectra. For instance the elongated helix of polyproline II has a maximum at about 205 nm. The ord spectrum of $\beta$-sheet structures reveals a trough at 205 nm and a peak at 190 nm. These relatively simple relationships are complicated, however, by the simultaneous presence of more than one kind of geometry in different parts of the same molecule and also by chromophores other than the amide group. Tyrosine, tryptophan and to some extent phenylalanine residues modify the spectra. Hence, the interpretation of ord and cd spectra is not always unequivocal and does not necessarily allow the definitive assignment of a certain architecture. Nevertheless, chiroptical spectra are of considerable value: they can demonstrate the presence or absence of a preferred conformation and can reveal the effect of organic solvents, detergents, changes in concentration or in temperature on the architecture of a peptide. It is also possible to compare with their help the influence of individual amino acid residues on the geometry through a study of synthetic analogs. The results of such investigations must of course be evaluated judiciously. For instance, because of the marked increase in helix content on addition of organic solvents or detergents to aqueous solutions of peptides, it is often claimed that the helical form is the biologically active conformation of a certain compound and this claim is supported by the argument that a similar conformational change has to be expected on contact with its specific receptor, usually a lipid-rich membrane. It is difficult to prove or to disprove such claims, since no isolated receptors are available for a direct study of peptide-receptor interaction. Yet, in spite of such reservations, the usefulness of chiroptical spectra remains uncontested. They can be recorded in dilute aqueous solutions, while nuclear magnetic resonance spectra usually require higher concentrations and x-ray crystallography is performed in the even less "physiological" solid state.

## 2. Nuclear Magnetic Resonance (nmr) Spectroscopy

The traditional methods of nmr spectroscopy have considerable value in structure determination and were widely used for corroboration of the identity of intermediates in peptide synthesis. A new era was opened up, however, with the development of *two dimensional nmr spectroscopy*. It is able to make signal assignment in peptides more productive. Thus determination of connectivity between spin systems within the same amino acid allows the identification of individual amino acid residues while homonuclear and heteronuclear couplings across the peptide bond

$$
\begin{array}{ccc}
\mathrm{R} & & \mathrm{R} \quad \mathrm{O} \\
| & & | \quad \| \\
-\mathrm{N}-\mathrm{C}- & & -\mathrm{C}-\mathrm{C}-\mathrm{N}- \\
| \quad | & & | \qquad | \\
\mathrm{H} \quad \mathrm{H} & & \mathrm{H} \qquad \mathrm{H}
\end{array}
$$

are used for the elucidation of sequence. The significance of nmr spectroscopy increased in a major way by the study of long range interactions with the help of various nmr techniques, such as nuclear Overhauser effect enhanced spectra, which assist the determination of through-space connectivities (Kessler 1982). Not less significant is the extension of nmr investigations to nuclei such as $C^{13}$, $O^{17}$ and $N^{15}$. With the help of advanced techniques nmr spectroscopy is now the perhaps most important tool in the conformational analysis of peptides. The results of x-ray crystallography remain important but they can be obtained only when suitable crystals are available and this is less frequently the case than one would expect. Also, the geometry determined through x-ray diffraction patterns may be valid only for the solid state or for the particular crystal form and not for conformation in solution. The principal advantage of nmr spectroscopy lies in the fact that the spectra recorded are those of peptides in solution. Thus the interactions that stabilize the architecture are mainly intramolecular. Intermolecular interaction, that might alter the conformation can be diminished by decreasing the concentration of the dissolved peptide molecules.

This impressive area of conformational analysis is still rapidly expanding and can not be rendered in this volume. A book written by Wüthrich and an article by Kessler and his associates, both listed at the end of this chapter, will satisfy the interested reader.

## 3. X-Ray Crystallography

Both the theoretical background of x-ray crystallography and its application for the elucidation of three dimensional structure of peptides transcend the boundaries of peptide chemistry. Some special articles written on this subject are cited at the end of this chapter to provide sources for those who wish to pursue this topic in more detail. Here we can merely comment on the scope and significance of the method.

At present no other experimental approach provides such detailed and exact information about the spatial arrangement of atoms in the molecule of a peptide as the solution of single crystal x-ray diffraction patterns. Two dimensional nmr spectroscopy, discussed in the preceding section, complements rather than displaces x-ray crystallography, because it is applied in solution and can be used therefore for compounds that do not afford nice single crystals. A surprisingly large number of peptides failed to crystallize so far. This might be due to the conformational freedom which, in spite of intramolecular interactions can still be found in the molecules of peptides. Proteins show a greater tendency for crystallization, probably because numerous intramolecular forces contribute to a well defined architecture. Thus, a globular protein can become a building component of a crystal lattice without undergoing a major conformational change. This is not true for certain peptides which exhibit substantially different geometries in solution and in the solid state. Also, the crystals that sometime do form are often only microscopic. They are suitable for powder-diagrams (Debye patterns) but not for single crystal studies. The latter require crystals with about 0.1 mm size in all three directions. The powder diagrams are not worthless, but allow only a rather general description of the architecture, such as "helical" or "$\beta$-sheet" structure. Only with single crystals can electron-density maps be secured which then allow calculation of the molecular geometry in fine detail. Recent advances in computation greatly extended the scope of such studies: it became possible to dispense with the formerly obligatory heavy atom substituent and isomorphic substitution. In several instances conformation of peptides emerged with such detail and clarity that it seemed that x-ray crystallography is the last word in conformational analysis for peptides. In reality, however, the alternative methods retain much of their value. Since the geometry of a peptide in the crystal is usually affected by packing forces that comprise also intermolecular interactions such as hydrogen bonds, dipole-dipole interactions, hydrophobic bonds or ion-pair formation, the conformational changes that take place during crystallization can be quite drastic. It is not unprecedented to find two different geometries in two crystalline forms of the same peptide or even within the same crystals. Hence, it seems to be mandatory to examine the general validity of the structure so determined through the application of alternative methods, mainly ord-cd spectra and two-dimensional nmr spectroscopy.

# C. Prediction of Conformation in Peptides

Interactions between side chains of neighboring amino acid residues can be calculated by several methods (Scheraga 1979) and it is possible, therefore, to predict from the sequence of amino acids the geometry of the peptide. Yet, extensive studies based for instance on potential energy-surface calculations were only moderately successful so far: the calculated geometry was contradicted by subsequently established experimental evidence in several cases. Such discrepancies would be less frequent if the effect of solvents and of long-range interactions could be included in the calculations. As discussed in previous sections, the conformation of peptides is mostly solvent dependent and can also change with concentration: at high dilution intermolecular interactions have less influence. Also, residues which are distant along the chain, when brought into proximity by folding have an important role in the stabilization of secondary structures. These complicating factors can not be ignored in the calculation of geometry and it might be more realistic at this time to use algorithms (Lim 1974) based on well established architectural features of globular proteins such as the presence of hydrophobic residues in tightly packed pockets, hydrophilic side chains on the surface of the molecule, etc.

A simple and popular method for the prediction of conformation is the application of empirical conformational parameters. From the study of numerous protein structures determined by x-ray crystallography it became obvious that certain amino acids are frequent constituents of helical portions while others are more likely to occur in $\beta$-sheets and still others are often found in reverse turns. Statistical treatment of such frequencies (Chou and Fasman 1978) produced sets of conformational parameters (Table 5) which were applied for the prediction of conformation in proteins with considerable success. A cluster of four helical residues ($P_\alpha > 1.00$) in a six residue stretch of a chain is predictive for a helical region that can be extended in both directions until a

**Table 5.** Chou-Fasman conformational parameters

|  | Ala | Arg | Asn | Asp | Cys | Gln | Glu | Gly | His | Ile | Leu | Lys | Met | Phe | Pro | Ser | Thr | Trp | Tyr | Val |
|---|---|---|---|---|---|---|---|---|---|---|---|---|---|---|---|---|---|---|---|---|
| $P_\alpha$ | 1.42 | 0.98 | 0.67 | 1.01 | 0.70 | 1.11 | 1.51 | 0.57 | 1.00 | 1.08 | 1.21 | 1.16 | 1.45 | 1.13 | 0.57 | 0.77 | 0.83 | 1.08 | 0.69 | 1.06 |
| $P_\beta$ | 0.83 | 0.93 | 0.89 | 0.54 | 1.19 | 1.10 | 0.37 | 0.75 | 0.87 | 1.60 | 1.30 | 0.74 | 1.05 | 1.38 | 0.55 | 0.75 | 1.19 | 1.37 | 1.47 | 1.70 |

$P_\alpha$ is the statistical probability for a residue to be in an $\alpha$-helical region

$P_\beta$ is the statistical probability for a residue to be in a $\beta$-sheet region

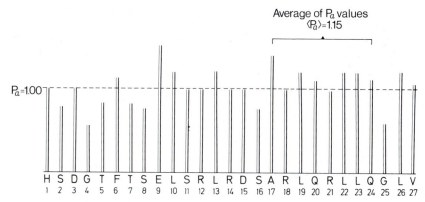

**Fig. 2.** $P_\alpha$ values for secretin

helix breaker ($P_\alpha < 1.00$) occurs. Proline can not be one of the residues in the helical portion of the chain. Prediction of an $\alpha$-helix requires that the average of the $P_\alpha$ values $\langle P_\alpha \rangle$ be more than 1.03 and higher than the average of the $\beta$-sheet parameters $\langle P_\beta \rangle$. For $\beta$-sheets the average of the individual parameters $\langle P_\beta \rangle$ must be higher than 1.05. Reverse turns are less easily predicted because residues preceding and following the turn also must be taken into consideration. It is interesting to note that proline is a frequent constituent of reverse turns ($P_t = 1.52$) and so is glycine ($P_t = 1.56$), while they rarely occur in helical regions or in $\beta$-sheets. For further details of the prediction rules the comprehensive article of Chou and Fasman should be consulted.

Since (with the exception of microbial products) biologically active peptides are formed through specific fragmentation of proteins, it seems to be justified to adopt empirical conformational parameters for the prediction of conformation of peptides as well. A histogram drawn for the gastrointestinal peptide secretin (Fig. 2) suggests an $\alpha$-helical stretch in the C-terminal half of the chain, encompassing residues 17 to 24. The calculations suggested two reverse turns, one near the N-terminus, the other at the center of the chain. There was no indication for the presence of a $\beta$-sheet. Experimental data support this prediction and also show the stabilization of the helical stretch by folding of the chain (Bodanszky and Bodanszky 1986).

# References

Bodanszky, M., Bodanszky, A.: Peptides *7*, 43 (1986)
Chou, P., Fasman, G. D.: Ann. Review Biochem. *1978*, 251
Förster, T.: Ann. Phys. *2*, 55 (1948)
Kauzmann, W.: Adv. Prot. Chem. *14*, 1 (1959)

Kessler, H.: Angew. Chem. Int. Ed. Engl. *21*, 512 (1982)

Lim, V. T.: J. Mol. Biol. *88*, 857, 873 (1974)

Pauling, L., Corey, R. B., Branson, H. R.: Proc. Nat. Acad. Sci. USA *37*, 205 (1951)

Ramachandran, G. N., Ramakrishnan, C., Sasisekharan, V.: J. Mol. Biol. *7*, 95 (1963)

Scheraga, H. A.: Acc. Chem. Res. *12*, 7 (1979)

Schiller, P. W. in: The Peptides Vol. 7 (Udenfriend, S., Meienhofer, J. eds.) Academic Press, Orlando, Fla 1985, p. 115

Venkatachalam, C. M.: Biopolymers *6*, 1425 (1968)

White, F. H.: J. Biol. Chem. *236*, 1353 (1961)

# Additional Sources

Ramachandran, G. N., Sasisekharan, V.: Conformation of Polypeptides and Proteins in Adv. Prot. Chem. *23*, 283–437 (1968)

Schulz, G. E., Schirmer, R. H.: Principles of Protein Structure, Springer Verlag, New York 1979

Herskovits, T. T.: Difference Spectroscopy, in Methods of Enzymology, Vol. 11 (Hirs, C. H. W. ed) Academic Press, New York 1967, pp. 870–905

White, F. H.: Reduction and Reoxidation of Disulfide Bonds, in Methods of Enzymology, Vol. 11 (Hirs, C. H. W. ed) Academic Press, New York 1967, pp. 481–485

Craig, L. C.: Techniques for the Study of Peptides and Proteins by Dialysis and Diffusion, in Methods of Enzuymology, Vol. 11 (Hirs, C. H. W. ed) Academic Press, New York 1967, pp. 870–895

Jirgenson, B.: Optical Rotatory Dispersion of Proteins and Other Macromolecules, Springer Verlag, New York 1969

Woody, R. W.: Circular Dichroism of Peptides, in The Peptides, Vol. 7 (Udenfriend, S., Meienhofer, J. eds) Academic Press, Orlando, Fla. 1985, pp. 15–114

Wüthrich, K.: NMR in Biological Research, Peptides and Proteins. North Holland. Publ. Amsterdam 1976

Kessler, H., Bermea, W., Müller, H., Pook, K. H.: Modern Nuclear Magnetic Resonance Spectroscopy, in The Peptides, Vol. 7 (Udenfriend, S., Meienhofer, J. eds) Academic Press, Orlando, Fla. 1985, pp. 437–473

Karle, I.: X-Ray Analysis: Conformation of Peptides in the Crystalline State in The Peptides, Vol. 4 (Gross, E., Meienhofer, J., eds) Academic Press, New York 1981, pp. 1–54

Benedetti, E.: Structure and Conformation of Peptides as Determined by x-Ray Crystallography, in Chemistry and Biochemistry of Amino Acids, Peptides and Proteins, Vol. 6 (Weinstein, B. ed) M. Dekker, New York 1982, pp. 105–184

Gunnin, J., Blundell, T.: Crystal Structure Analysis of Larger Peptide Hormones, in The Peptides, Vol. 4 (Gross, E., Meienhofer, J. eds) Academic Press, New York 1981, pp. 55–84

Zimmerman, S. S.: Theoretical Methods in the Analysis of Peptide Conformation, in The Peptides, Vol. 7 (Udenfriend, S., Meienhofer, J. eds) Academic Press, Orlando, Fla. 1985, pp. 165–212

# Part Two: Peptide Synthesis

Synthesis of a peptide with a well defined sequence of amino acid residues is a fairly complex process which appears simple only for those who were never involved in it. This complexity does not follow from the construction of the peptide bond: numerous methods are available for that task, most of them quite good, and the only problem in this respect is to select the one most suited for the actual bond under consideration. A more demanding part of the discipline of peptide synthesis is the necessity to block those functional groups which should not participate in the peptide bond forming reaction or as usually called "coupling". Blocking groups (or protecting groups or masking groups) are needed for the amino group of the amino acid which lends its (activated) carboxyl group to the coupling reaction (the "carboxyl component") and also for the carboxyl group of the amino acid to be acylated at its amino group (the "amine component"). In order to lengthen the resulting dipeptide derivative one of the masking groups has to be removed and this should be possible without harm to the peptide bond or to the "semipermanent" protecting group which has to stay in place until a later stage when its cleavage becomes necessary. Not lastly, the various functional groups on amino acid side chains also require protection. Some of these maskings are mandatory, others only optional. Aspects of coupling and protection will be discussed in Chapters V and VI respectively, followed by a chapter presenting numerous undesired side reactions that can occur during these operations. Chapter VIII deals with a special side reaction, central to the methodology of peptide synthesis, namely racemization. Tactics and strategies designed for the conservation of chiral purity and synthesis of special target compounds, such as poly-amino acids, sequential peptides, cyclic peptides, etc. are discussed in Chapter IX. A somewhat detailed treatment of solid phase peptide synthesis in Chapter X is followed by the brief presentation of various other methods of facilitation (Chapter XI). The second part of the volume is concluded with an overview of analysis and characterization of synthetic peptides, mainly for the sake of those who wish to become involved in the actual execution of peptide synthesis.

# V. Formation of the Peptide Bond

In order to convert carboxylic acids into acylating agents their hydroxyl group must be replaced by an electron-withdrawing substituent (X) to enhance the polarization of the carbonyl group and thereby the electrophilicity of its carbon atom. Thus the nucleophilic attack by the amino group (of the amino acid to be acylated) is greatly facilitated:

$$\text{Activation:} \quad R-COOH \longrightarrow R-\overset{\displaystyle O}{\overset{\|}{C}}-X$$

Coupling:

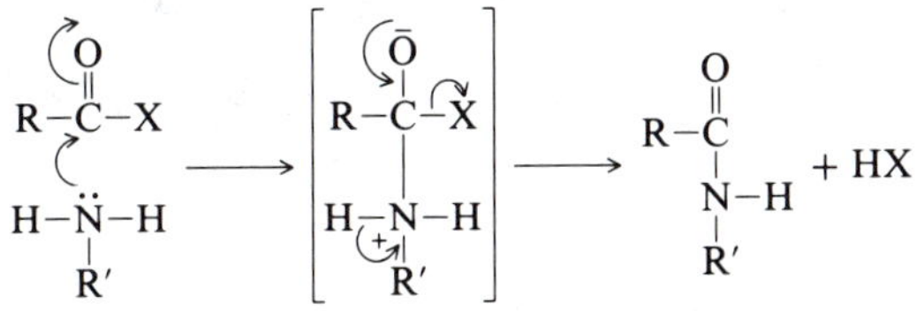

## A. The Acid Chloride Method

The chlorine atom is an obvious choice for the role of the electron-withdrawing moiety (X). The powerful activation present in acid chlorides was well known by the turn of the century when the first attempts at peptide bond formation were made. Yet, both preparation and application of amino acid chlorides are far from unequivocal and the success achieved with the acid chloride method (Emil Fischer 1903) remained fairly limited. Conversion of the carboxyl group of the amino acid (partially blocked by the amine-protecting group Y) to the chloride was carried out with phosphorus pentachloride in the cold:

$$Y-NH-CHR-COOH + PCl_5 \longrightarrow Y-NH-CHR-\overset{\displaystyle O}{\overset{\|}{C}}-Cl + HCl + POCl_3$$

While the HCl generated in the reaction readily escapes, the acid chlorides remain contaminated with the less volatile phosphorus oxychloride. The use of thionyl chloride for acid chloride formation would seem to be an effective remedy since both by-products, HCl and $SO_2$, are volatile:

$$R-COOH + SOCl_2 \longrightarrow R-\overset{\overset{O}{\|}}{C}-Cl + SO_2 + HCl$$

Under the conditions of the reaction, however, the commonly applied alkyloxy-carbonyl blocking groups become involved in the process and give rise to the formation of N-carboxyanhydrides (Leuchs anhydrides; Leuchs 1906)

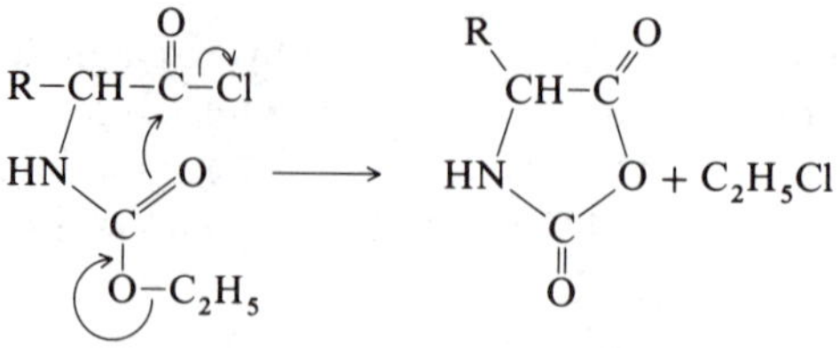

which are reactive derivatives of amino acids, suitable for acylation but not for the general synthesis of well defined sequences. Also, intramolecular nucleophilic attack in acylamino acid chlorides yields "azlactones"

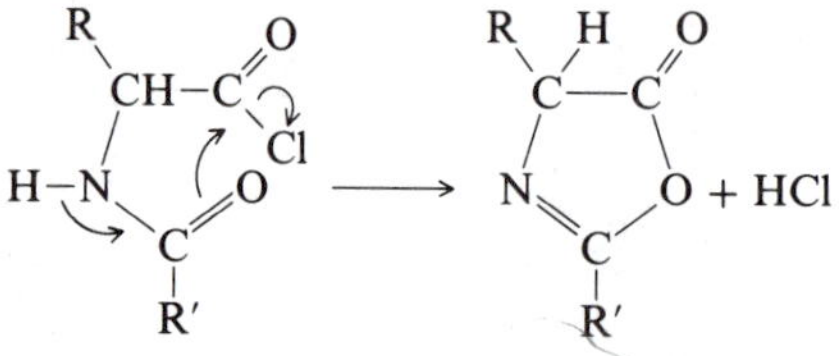

derivatives of dihydro-oxazolone, which are good acylating agents but were often implicated in racemization. There were several attempts for the revival of the acid chloride method, for instance by the application of more sophisticated reagents such as oxalyl chloride or N,N-dimethylformamidium chloride (from dimethyl formamide and phosgene)

$$(CH_3)_2N{=}CHCl^+ \cdot Cl^-$$

but without major success so far. Acceptable results were achieved when the blocking group was resistant to thionyl chloride, for instance in the case of p-toluenesulfonylamino acids. It is reasonable to expect similarly successful activation of amino acids blocked by the acid resistant 9-fluorenylmethyloxy-carbonyl (Fmoc) group.


## B. The Acid Azide Method

It is interesting to note that activation in the form of acid azides (Curtius 1902), the sole competitor of the acid chloride procedure at the time of its discovery, remains until today a powerful and practical approach for the synthesis of peptides. The original steps of activation, to wit, hydrazinolysis of alkyl esters and conversion of the hydrazides to acid azides with the help of nitrous acid

$$R-\overset{\overset{\displaystyle O}{\|}}{C}-OCH_3 \xrightarrow[(-CH_3OH)]{H_2NNH_2} R-\overset{\overset{\displaystyle O}{\|}}{C}-NHNH_2 \xrightarrow[(-2H_2O)]{HONO} R-\overset{\overset{\displaystyle O}{\|}}{C}-N=\overset{+}{N}=\overset{-}{N}$$

are still practiced, although direct conversion of carboxylic acids to acid azides with the help of diphenylphosphoryl azide

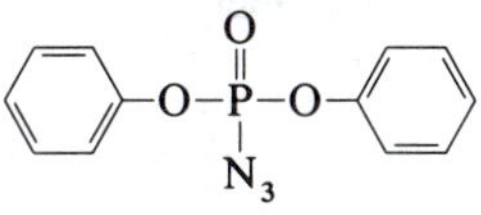

became a viable alternative in recent years. Also, instead of nitrous acid, alkyl nitrites can be used, in fact sometimes with improved results, for the transformation of hydrazides to azides.

$$R-\overset{\overset{\displaystyle O}{\|}}{C}-NHNH_2 + C_4H_9ONO \longrightarrow R-\overset{\overset{\displaystyle O}{\|}}{C}-N_3 + C_4H_9OH + H_2O$$

These improvements, however, did not effect an inherent shortcoming of the azide method, the well studied Curtius rearrangement of acid azides to isocyanates:

$$R-\overset{\overset{\displaystyle O}{\|}}{C}-N=\overset{+}{N}=\overset{-}{N} \longrightarrow R-N=C=O + N_2$$

The latter react with the amine component and produce urea derivatives

$$R-N=C=O + H_2N-R' \longrightarrow R-NH-CO-NH-R'$$

which are rather similar in their properties to the desired peptide and can be separated from it only with difficulty. Because of this important side reaction it is imperative to use all possible precautions in the execution of the azide process, such as low temperature and high concentration of the reactants. Yet, in spite of these problems, the azide process remains one of the truly valuable assets of peptide chemists, mainly because in most cases coupling via azides is accompanied with only negligible racemization. Even this brief treatment of the azide method can not be concluded without a word of warning about the toxic nature of hydrazoic acid produced during acylation:

$$R-\overset{\overset{\displaystyle O}{\|}}{C}-N=\overset{+}{N}=\overset{-}{N} + H_2N-R' \longrightarrow R-CO-NH-R' + HN_3$$

## C. Anhydrides

One of the simplest and most efficient methods of acylation is the treatment of amines (or other nucleophiles) with anhydrides of carboxylic acids. Acetylation with acetic anhydride is common practice and it may seem rather surprising that the analogous reaction has not been used for peptide bond formation from the earliest days of peptide synthesis. One possible explanation for this delay in the acceptance of anhydrides is the wastefulness of the process. Of the two molecules of the carboxylic acid comprised in the molecule of the anhydride only one is incorporated into the product while the other is regenerated but usually not recovered:

$$R-\overset{\overset{O}{\|}}{C}-O-\overset{\overset{O}{\|}}{C}-R + H_2N-R' \longrightarrow R-CO-NH-R' + R-COOH$$

The loss of half of the carboxylic acid is no cause for concern in acetylation with acetic anhydride, but appeared to be prohibitive when blocked amino acids, obtained sometimes with considerable effort, had to be sacrificed. This situation changed with time. Protected amino acids are now prepared with increasing facility and decreasing cost and their anhydrides are frequently applied. An interesting but not widely practiced approach to *symmetrical anhydrides* is the use of phosgene:

$$2R-COOH + COCl_2 \xrightarrow{-HCl} R-\overset{\overset{O}{\|}}{C}-O-\overset{\overset{O}{\|}}{C}-O-\overset{\overset{O}{\|}}{C}-R \longrightarrow$$

$$R-\overset{\overset{O}{\|}}{C}-O-\overset{\overset{O}{\|}}{C}-R + CO_2$$

A more attractive method, the application of condensing agents particularly carbodiimides,

$$2R-COOH + R'-N=C=N-R' \longrightarrow R-\overset{\overset{O}{\|}}{C}-O-\overset{\overset{O}{\|}}{C}-R + R'-NH-\overset{\overset{O}{\|}}{C}-NH-R'$$

will be discussed in more detail in connection with coupling reagents.

Formation of the peptide bond via *mixed* or *unsymmetrical anhydrides* gained considerable popularity after their introduction in the early forties. These intermediates were constructed in such a way that the blocked amino acid is practically the sole acylating moiety while the acid which plays the role of activator is eliminated. Acetic acid or benzoic acid are not well suited for this role. A search for better solutions led to branched long-chain fatty acids, such as isovaleric acid (Vaughan and Osato 1951). In these mixed anhydrides (prepared via the acid chlorides) reactivity of the activating part of the anhydride molecule is decreased both by electron release and by steric hindrance:

$$R-COO^- + Cl-\overset{O}{\overset{\|}{C}}-CH_2CH(CH_3)_2 \longrightarrow R-\overset{O}{\overset{\|}{C}}-O-\overset{O}{\overset{\|}{C}}-CH_2CH(CH_3)_2 + Cl^-$$

$$R-\overset{O}{\overset{\|}{C}}-O-\overset{O}{\overset{\|}{C}}-CH_2CH(CH_3)_2 + H_2N-R' \longrightarrow R-CO-NH-R'$$
$$+ (CH_3)_2CHCH_2COOH$$

Accordingly, the attack by the nucleophilic amine component will occur mainly at the carbonyl group of the blocked amino acid and the second possible acylation product, an isovalerylamine,

$$(CH_3)_2CHCH_2CO-NH-R'$$

is produced in very small, most often negligible amount. A further enhancement of this kind of steric-electronic effect is found in trimethylacetic acid (pivalic acid) mixed anhydrides:

$$R-\overset{O}{\overset{\|}{C}}-O-\overset{O}{\overset{\|}{C}}-C(CH_3)_3$$

The most successful realization of the mixed anhydride concept is, however, the generation of reactive intermediates with the help of alkyl chlorocarbonates (also called alkyl chloroformates):

$$R-COO^- + Cl-\overset{O}{\overset{\|}{C}}-OR' \longrightarrow R-\overset{O}{\overset{\|}{C}}-O-\overset{O}{\overset{\|}{C}}-OR' + Cl^-$$

Ethylcarbonic acid mixed anhydrides (Boissonnas 1951; Wieland and Bernhard 1951) were readily accepted in the practice of peptide synthesis and together with their improved version, isobutylcarbonic acid mixed anhydrides (Vaughan 1951) remain highly valued tools of peptide chemists. A particular advantage of alkylcarbonic acid mixed anhydrides is that the by-products of the acylation reaction, carbon dioxide and the appropriate alcohol, are easily removed from the mixture:

$$R-\overset{O}{\overset{\|}{C}}-O-\overset{O}{\overset{\|}{C}}OCH_2CH(CH_3)_2 + H_2N-R' \longrightarrow$$
$$R-CO-NH-R' + CO_2 + (CH_3)_2CHCH_2OH$$

In the application of mixed anhydrides a small amount of the second acylation product must be expected. No such complication is associated with the use of symmetrical anhydrides. Thus, in spite of wide experimentation with various mixed anhydrides, at this time only the isobutylcarbonic acid mixed anhydride procedure remains a viable competitor of the symmetrical anhydride method.

## D. Active Esters

The inherent ambiguity in coupling via mixed anhydrides can be avoided if aminolysis of esters is used for peptide bond formation. It is a well established practice to prepare amides of blocked amino acids or peptides by treating their alkyl esters with ammonia:

$$Y-NH-CHR-CO-NH-CHR'-CO-NH-CHR''-\overset{\overset{O}{\|}}{C}-OCH_3 + NH_3 \longrightarrow$$

$$Y-NH-CHR-CO-NH-CHR'-CO-NH-CHR''-CONH_2 + CH_3OH$$

Since in the ester only one electrophilic center is present no second acylation product can be generated. The modest reacitivty of methyl esters is compensated by the use of ammonia in large excess but completion of the ammonolysis reaction still requires considerable time. Because it is impractical to use amine components in large excess, this approach can not be applied for the formation of an amide bond between two amino acid residues, at least not without major improvements. While the reactivity of the amino group could not be enhanced so far, the ester group appeared to be well suited for modifications. A negative inductive effect in the alcohol component of the ester renders the carbon atom of the carbonyl group more electrophilic and hence it becomes more ready to accept the attack of nucleophiles, such as the amino group of the amine component. Thus, methyl esters could be activated by substitution with strongly electron-withdrawing groups, for instance with the cyano group. Aminolysis of cyanomethyl esters (Schwyzer et al. 1955) was found indeed suitable for the preparation of peptides

$$R-\overset{\overset{O}{\|}}{C}-OCH_2CN + H_2N-R' \longrightarrow R-CO-NH-R' + HOCH_2CN$$

but satisfactory rates were observed only when the reactants were applied in high concentration. Therefore, the method could be used for the synthesis of dipeptides and, to some extent, for the preparation of short peptide chains as well, but for a more general approach to peptide bond formation obviously a further increase in the reactivity of the ester group was necessary. *Aryl esters* proved to be more promising and enhancement of their inherent reactivity by electron-withdrawing substituents yielded a whole series of useful acylating agents. Thus, p-nitrophenyl esters (Bodanszky 1955) are still among the tools of peptide chemists. The even higher activation present in *ortho*-nitrophenyl esters

$$R-\overset{\overset{O}{\|}}{C}-O-\langle\text{C}_6\text{H}_4\rangle-NO_2 + H_2N-R' \longrightarrow R-CO-NH-R' + HO-\langle\text{C}_6\text{H}_4\rangle-NO_2$$

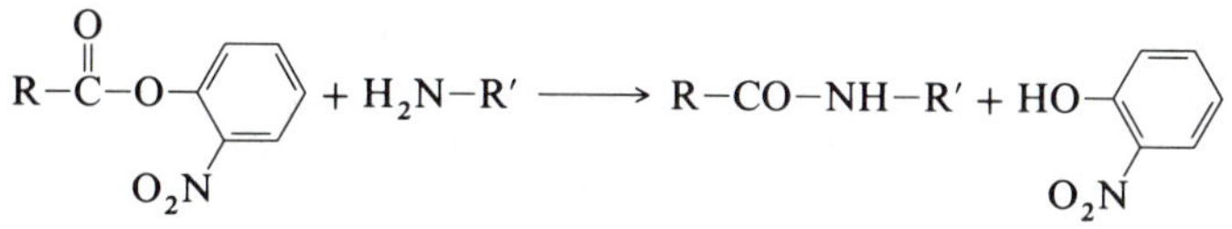

$$R-\overset{\overset{\displaystyle O}{\|}}{C}-O-\underset{O_2N}{\langle\bigcirc\rangle} + H_2N-R' \longrightarrow R-CO-NH-R' + HO-\underset{O_2N}{\langle\bigcirc\rangle}$$

(Bodanszky et al. 1969) is accompanied by diminished sensitivity to steric hindrance and also by decreased dependence of the reaction rates from the nature of the solvent used in the coupling step. Para-nitrophenyl esters are most active in highly polar media, such as dimethylformamide or dimethylsulfoxide, less active in ethyl acetate, dioxane or tetrahydrofurane and rather inert in methylene chloride or chloroform. In spite of these advantages mainly the para derivatives are used, probably because a major improvement in peptide synthesis rendered these differences less important, to wit, the discovery of the catalytic effect of 1-hydroxybenzotriazole (König and Geiger 1973) on aminolysis of active esters. This effect could be rationalized by the assumption of a ternary complex between active ester, amine and catalyst

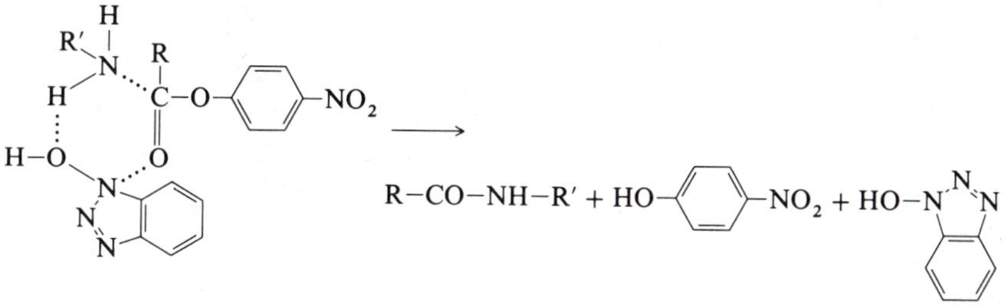

$$R-CO-NH-R' + HO-\langle\bigcirc\rangle-NO_2 + HO-N\overset{N}{\underset{N}{\diagdown}}N$$

but further rate-enhancement on addition of tertiary amines suggests that, at least in part, base catalyzed transesterification

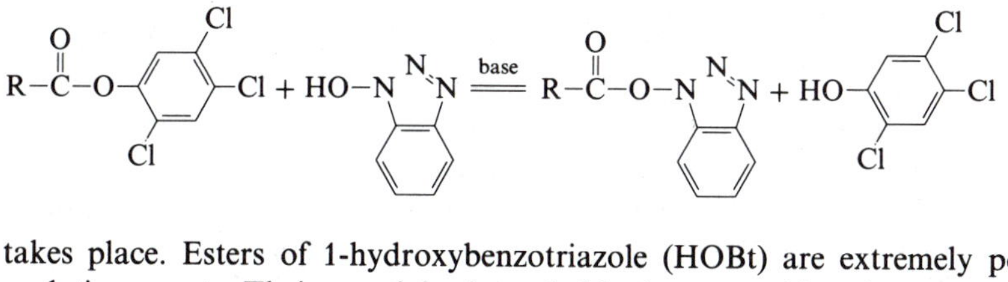

$$R-\overset{\overset{\displaystyle O}{\|}}{C}-O-\langle\bigcirc\rangle-Cl + HO-N\diagdown N \overset{base}{=\!=\!=} R-\overset{\overset{\displaystyle O}{\|}}{C}-O-N\diagdown N + HO-\langle\bigcirc\rangle-Cl$$

takes place. Esters of 1-hydroxybenzotriazole (HOBt) are extremely potent acylating agents. Their reactivity is probably due to anchimeric assistance:

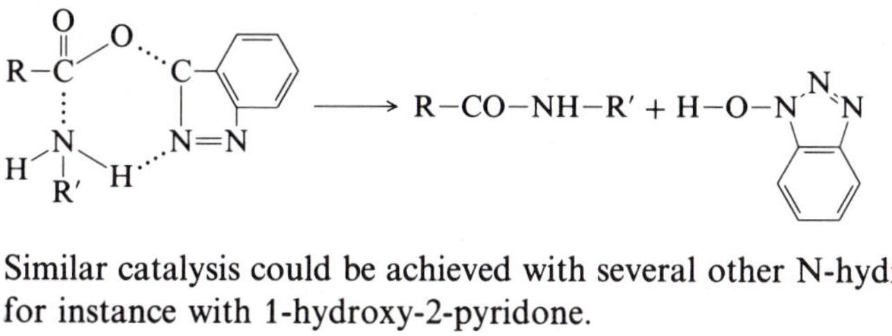

$$\longrightarrow R-CO-NH-R' + H-O-N\overset{N}{\diagdown}N$$

Similar catalysis could be achieved with several other N-hydroxy compounds, for instance with 1-hydroxy-2-pyridone.

Efficient syntheses of biologically active peptides, such as oxytocin or se-cretin, entirely by the nitrophenyl ester method stimulated further efforts to-ward the development of still more reactive esters but from the numerous activated intermediates proposed only the p-nitrophenyl esters and the esters of N-hydroxysuccinimide (Anderson et al. 1963) are commercially available at this time. The N-hydroxysuccimide esters are highly reactive and yield the desired peptides with the expected ease

$$R-\overset{O}{\overset{\|}{C}}-O-N\overset{\overset{O}{\overset{\|}{C}}}{\underset{\underset{O}{\overset{\|}{C}}}{\Big]}} + H_2N-R' \longrightarrow R-CO-NH-R' + HO-N\overset{\overset{O}{\overset{\|}{C}}}{\underset{\underset{O}{\overset{\|}{C}}}{\Big]}}$$

but their application is sometimes less than unequivocal: the succinimide car-bonyls are not entirely inert toward the amine component. The convenience found in the removal of the water soluble by-product, N-hydroxysuccinimide is an attractive feature of the method.

The exceptional reactivity of N-hydroxysuccimide and 1-hydroxybenzotria-zole esters is due to the effect of the neighboring N-atom. Similar kind of activation can be recognized in esters of 2-hydroxypyridine and 2-mercaptopy-ridine. A close-by oxygen atom, such as in catechol esters

$$R-\overset{O}{\overset{\|}{C}}-O-\overset{N=}{\langle\ \rangle} \qquad R-\overset{O}{\overset{\|}{C}}-S-\overset{N-}{\langle\ \rangle} \qquad R-\overset{O}{\overset{\|}{C}}-O-\underset{HO}{\langle\ \rangle}$$

is less effective in this respect.

For the preparation of amino acid polymers pentachlorophenyl esters (Kupryszewski 1961) have been recommended, because high molecular weight materials form preferably with them. Under the more general conditions of stepwise chain-building, however, the bulkiness of the activating group might outbalance the high degree of activation provided by pentachlorophenyl esters. This is especially true when steric hindrance is already a limiting factor, as in the incorporation of valine and isoleucine, or in solid phase peptide synthesis (conf. Chapter X) where the amino group is surrounded by the matrix of the polymeric support. This problem is less pronounced in the powerful pentafluo-rophenyl esters (Kova'cs et al. 1967):

$$R-\overset{O}{\overset{\|}{C}}-O-\underset{\underset{Cl\ \ \ Cl}{\overset{Cl\ \ \ Cl}{}}}{\langle\ \rangle}-Cl \qquad R-\overset{O}{\overset{\|}{C}}-O-\underset{\underset{F\ \ \ F}{\overset{F\ \ \ F}{}}}{\langle\ \rangle}-F$$

A detailed study (Pless and Boissonnas 1963) of the influence of halogen substituents on the reactivity of aryl esters led to the choice of 2,4,5-trichlorophenyl esters.

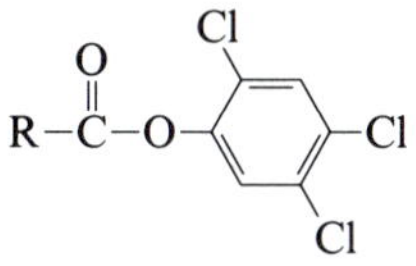

The same study revealed the unfavorable effect of substitution at *both* carbon atoms which are ortho to the phenolic hydroxyl. Hence, the so far untested 2,3,4,5-tetrachlorophenyl and 2,3,4,5-tetrafluorophenyl esters

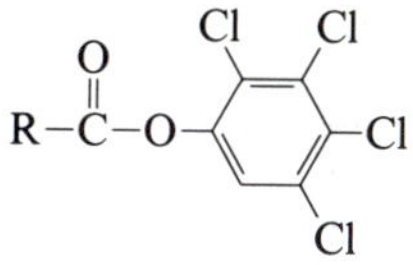

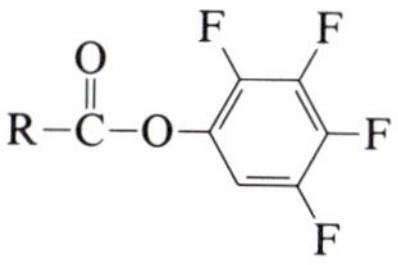

could bring about further improvement in the development of the active ester idea.


# E. Coupling Reagents

The introduction of carbodiimides (Sheehan and Hess 1955), particularly dicyclohexylcarbodiimide (DCC, DCCI) as reagents for the formation of the peptide bond was a major event in the history of peptide synthesis. The novel feature of coupling reagents was that they could be added to the *mixture* of the carboxyl component and the amine component. Thus, activation and coupling proceed concurrently. While amines do react with carbodiimides (yielding guanidine derivatives) the rate of this reaction is negligible when compared with the rapid rate observed in the addition of carboxylic acids to one of the double bonds of a carbodiimide:

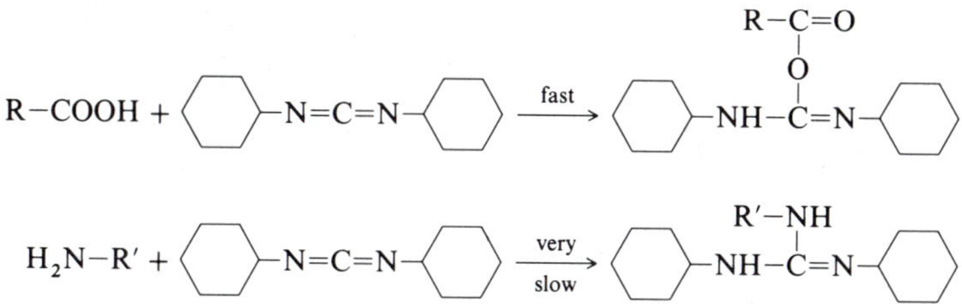

In O-acyl-isoureas, the intermediates formed in the addition of carboxylic acids to carbodiimides, the N = C group provides powerful activation which leads to coupling:

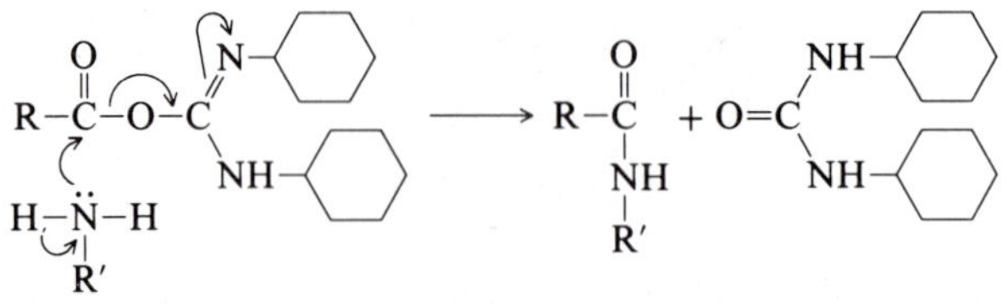

It seems to be reasonable to attribute some basic character [1] to O-acyl-isoureas and therefore general base catalysis

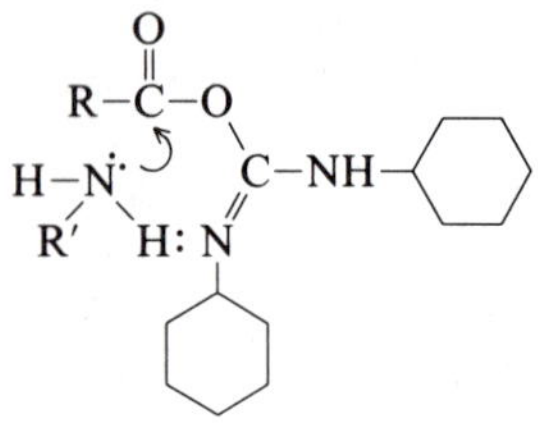

can be invoked as explanation of the surprisingly high reactivity observed in aminolysis.

A second pathway in the peptide bond forming reaction proceeds via symmetrical anhydrides produced in the attack of a yet unreacted molecule of the

---

[1] This assumption is based on analogies. For instance O-aryl isouronium salts are produced in the reaction of acidic phenols with carbodiimides:

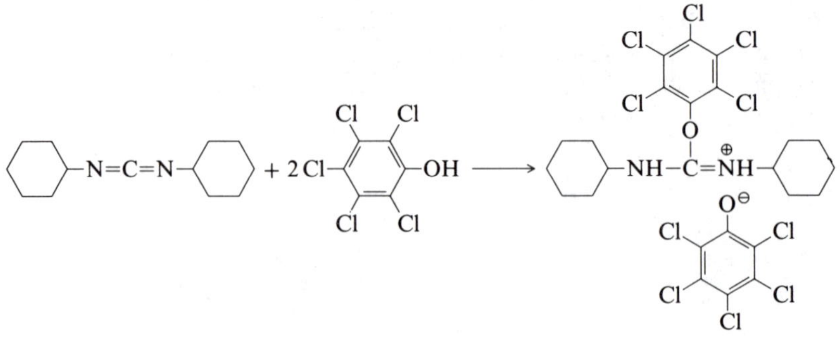

Also, it is well known that O-alkylisoureas are strong bases and even urea forms salts with strong acids.

carboxyl component on the O-acyl-isourea intermediate:

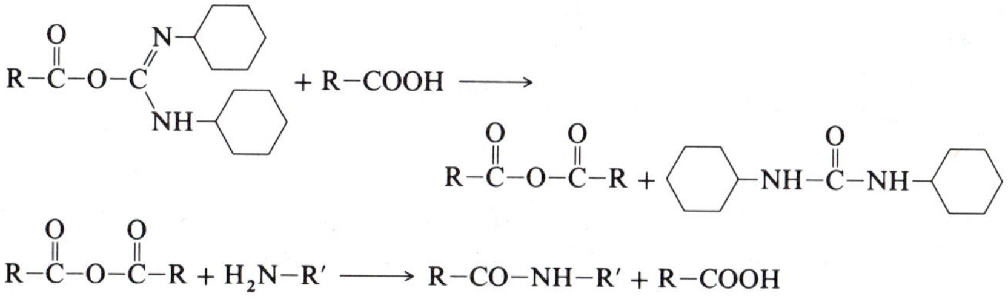

$$R-\overset{\overset{\text{O}}{\|}}{C}-O-\overset{\overset{\text{N}-}{}}{\underset{\text{NH}-}{C}} + R-COOH \longrightarrow$$

$$R-\overset{\overset{\text{O}}{\|}}{C}-O-\overset{\overset{\text{O}}{\|}}{C}-R + \overset{}{\underset{}{}}-NH-\overset{\overset{\text{O}}{\|}}{C}-NH-$$

$$R-\overset{\overset{\text{O}}{\|}}{C}-O-\overset{\overset{\text{O}}{\|}}{C}-R + H_2N-R' \longrightarrow R-CO-NH-R' + R-COOH$$

The extremely rapid generation of symmetrical anhydrides can be rationalized by the assumption of a quasi-intramolecular attack of the carboxylate anion on the reactive carbonyl within the ion-pair of the two components:

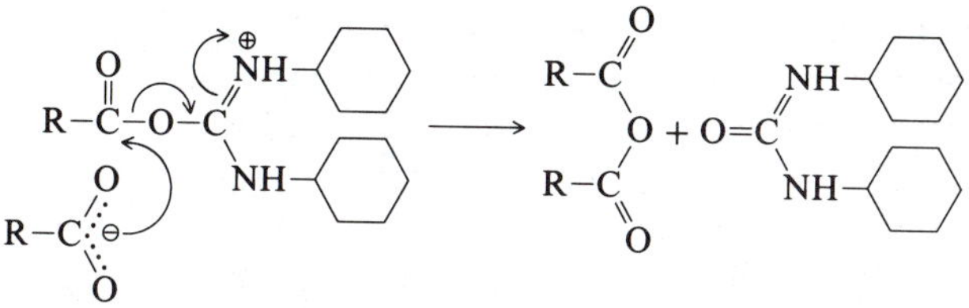

The speedy execution of activation and coupling in a single operation and the simple removal of the insoluble by-product, N,N'-dicyclohexylurea (DCU), by filtration all contributed to the persistent popularity of the DCC method. Not surprisingly a certain price has to be paid for expediency. Thus, overactivation in the reactive intermediate results in some loss of chiral purity when peptide segments are coupled. Furthermore, the nucleophilic center on O-acylisoureas competes with the amine component for the acyl residue and this competition leads to the formation of unreactive by-products, N-acylureas:

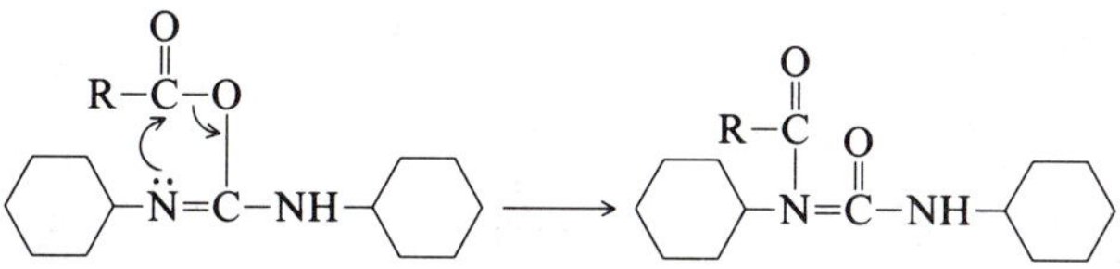

Both racemization and N-acylurea formation can be suppressed by the addition of auxiliary nucleophiles such as 1-hydroxybenzotriazole (HOBt) proposed by König and Geiger (1970). Attack of the additive on the reactive intermediate yields an O-acyl-1-hydroxybenzotriazole, a powerful acylating agent:

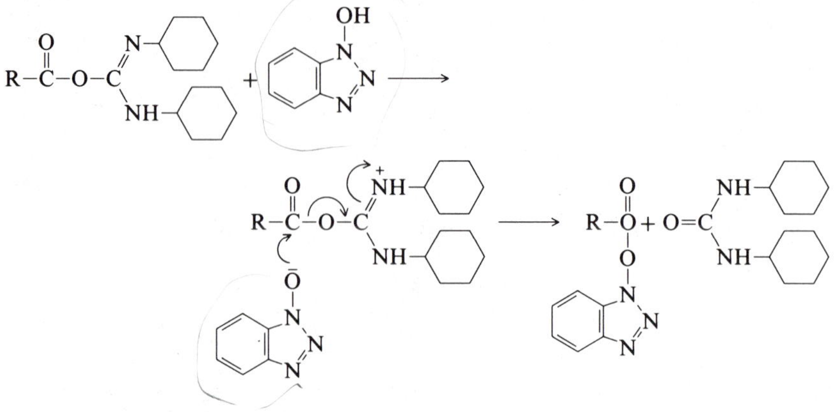

The presence of a second nucleophile in the reaction mixture reduces the concentration of the O-acylisourea and thereby the extent of racemization. Also, HOBt, a weak acid, prevents proton abstraction from the chiral carbon atom and thus contributes to the conservation of chiral purity in a second manner as well. Last, but not least the availability of the auxiliary nucleophile (HOBt) efficiently shortens the lifetime of the overactivated O-acyl-isourea intermediate and thus diminishes the extent of O → N acyl-migration leading to N-acylureas. It should be noted that HOBt is regenerated during acylation, hence its concentration remains almost constant during coupling.

The success achieved with dicyclohexylcarbodiimide stimulated an unrelenting search for even better coupling reagents. Some really effective compounds could be found, for instance carbonyldiimidazole (Staab 1957) which mediates coupling in a novel manner, through reactive N-acyl intermediates:

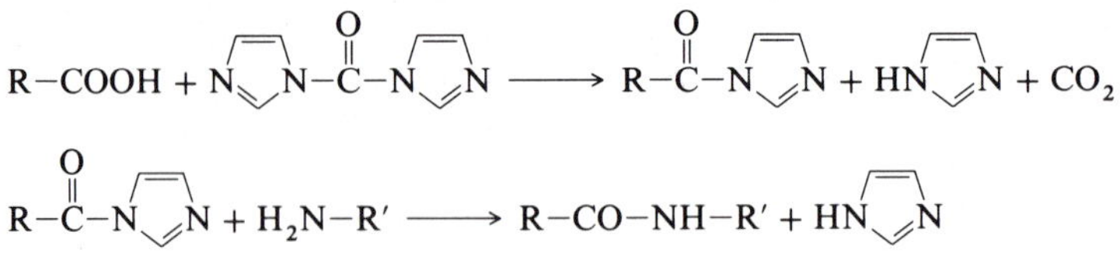

Unfortunately both in the preparation of the reagent and its application for peptide bond formation rigorously anhydrous conditions must be maintained and this appears to be too demanding for general use. A more convenient reagent, 1-ethyloxycarbonyl-2-ethyloxy-1,2-dihydroquinoline (EEDQ, Belleau and Malek 1968) and its improved version 1-isobutyloxycarbonyl-2-isobutyloxy-1,2-dihydroquinoline (IIDQ) generate the already discussed alkyloxycarbonic acid mixed anhydrides

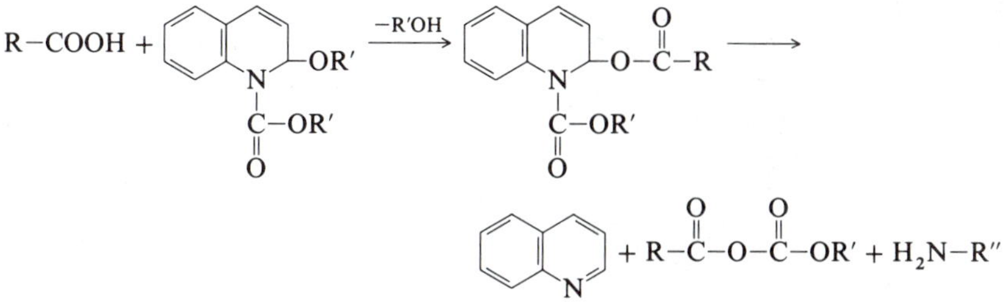

$$R-COOH + \underset{\underset{O}{\underset{\|}{C-OR'}}}{\overset{OR'}{\quad}} \xrightarrow{-R'OH} \underset{\underset{O}{\underset{\|}{C-OR'}}}{\overset{O-\overset{O}{\overset{\|}{C}}-R}{\quad}} \longrightarrow$$

$$+ R-\overset{O}{\overset{\|}{C}}-O-\overset{O}{\overset{\|}{C}}-OR' + H_2N-R''$$

$$R-\overset{O}{\overset{\|}{C}}-O-\overset{O}{\overset{\|}{C}}-OR' + H_2N-R'' \longrightarrow R-CO-NH-R'' + R'OH + CO_2$$

yet offer some advantages in the execution of the coupling. For instance no tertiary amine is needed to neutralize the HCl formed when mixed anhydrides are prepared via chlorocarbonates and quinoline, released during activation, is an extremely weak base that does not cause racemization or other side reactions by proton abstraction.

In more recent years phosphonium derivatives such as Bates' reagent (Bates et al. 1975)

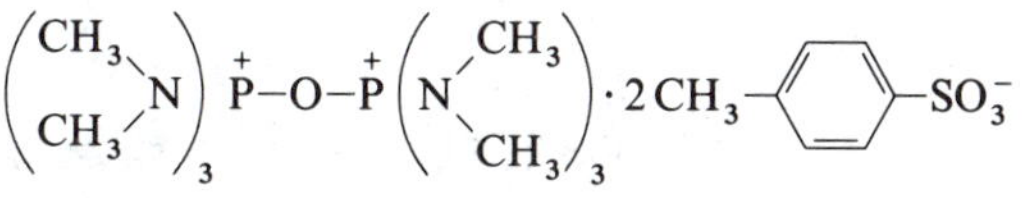

$$\left(\underset{CH_3}{\overset{CH_3}{\diagdown}}N\right)_3 \overset{+}{P}-O-\overset{+}{P}\left(N\underset{CH_3}{\overset{CH_3}{\diagup}}\right)_3 \cdot 2CH_3-\!\!\left\langle\phantom{x}\right\rangle\!\!-SO_3^-$$

or the BOP reagent, benzotriazolyl N-oxytri-dimethylamino-phosphonium hexafluorophosphate (Castro et al. 1975)

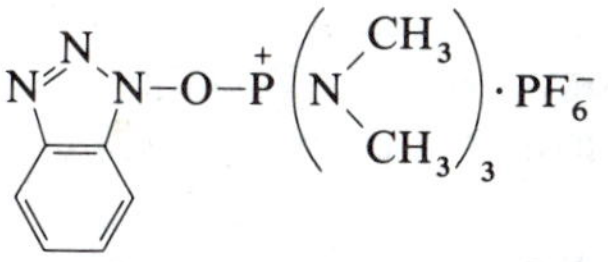

$$\underset{}{\overset{N}{\underset{N}{\diagup}}}\!\!N-O-\overset{+}{P}\left(N\underset{CH_3}{\overset{CH_3}{\diagup}}\right)_3 \cdot PF_6^-$$

found application in the synthesis of fairly complex peptides, but none of these compounds can yet compete in popularity with DCC.

At this point one must question the justification of the predilection for coupling reagents. The criteria of the perfect reagent are rather stringent: it should be inert toward the amine component (already present during activation in the reaction mixture), it should not generate a reactive intermediate containing a nucleophilic center (because this can compete with the amine component for the acyl group) and it should not cause overactivation which would lead to side reactions and hence to by-products. It is less demanding to activate the carboxyl group in the *absence* of the amine component, that is to separate the steps of activation and coupling. In fact, even the most successful reagent, DCC, is not entirely free from shortcomings and gives better results when used in activating mode and the symmetrical anhydrides thus produced are applied for the desired acylation.

## F. Enzyme Catalyzed Bond Formation

In the absence of acids or bases peptide bonds are quite resistant to hydrolysis, but their hydrolytic cleavage is extremely accelerated in the presence of proteolytic enzymes. The remarkable catalytic effect of these enzymes tempted many investigators, through a long period of time (Fruton 1982), to adopt them for synthesis, rather than hydrolysis of peptide bonds. Since enzymes are catalysts and merely accelerate the establishment of equilibria, it is possible to use proteolytic enzymes for amide bond formation if the equilibrium of the reaction can be modified. Thus anilides of blocked amino acids could be prepared with the help of papain:

$$Y-NH-CHR-COOH + H_2N- \bigcirc \xrightleftharpoons{papain}$$
$$Y-NH-CHR-CO-NH- \bigcirc + H_2O$$

Aniline was used in large excess and the anilide, being insoluble in water separates from the reaction mixture. Both factors shift the equilibrium to the right and therefore the anilide could be obtained in high yield. A practical application of this approach is the resolution of an enantiomeric mixture: only the L-derivative (of an N-acyl-amino acid) is converted, the D-amino acid derivative remains unchanged and can easily be separated from the insoluble anilide of the L-compound.

A more general use of proteolytic enzymes in peptide synthesis became feasible with the discovery (Sealock and Laskowsky 1969) of the effect of water miscible organic solvents on the equilibrium in enzyme catalyzed peptide bond hydrolysis and synthesis. In the presence of isopropanol (or dimethylformamide, etc.) the dissociation of the carboxyl group is suppressed and, at least in a selected pH region, the equilibrium is shifted toward synthesis. A notable case is the conversion of porcine insulin to human insulin. Enzymatic cleavage of the C-terminal residue of the B-chain (alanine) with carboxypeptidase yields desalanino pork insulin. This cleavage is followed by the incorporation of

L-threonine, in the form of its tertiary butyl ester, used in large excess and in the presence of isopropanol:

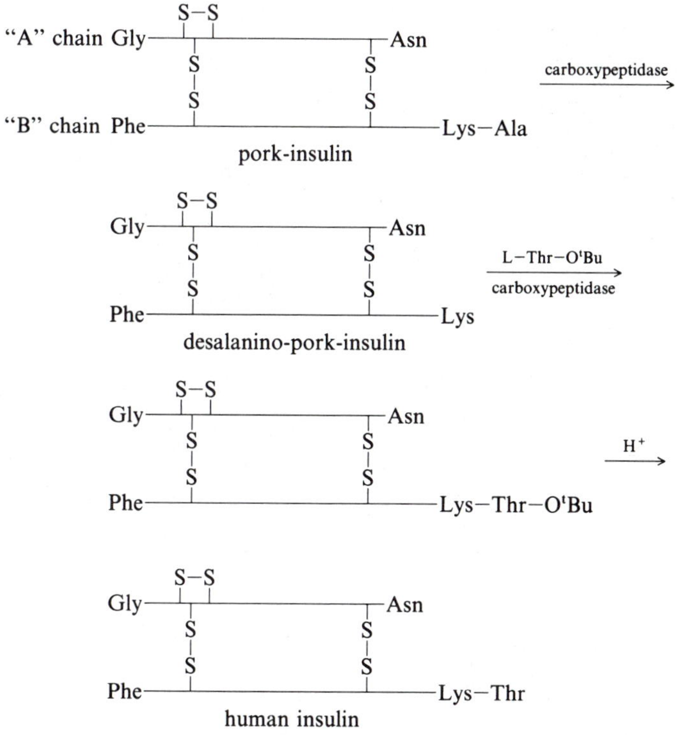

Acidolytic removal of the tertiary butyl group completes the conversion. This transformation is being carried out on a commercial scale.

In recent years numerous attempts were made to apply enzymes for the systematic building of peptide chains and some of these syntheses were fairly successfull. For instance the artificial sweetener, aspartame could be prepared without blocking the β-carboxyl group of aspartic acid. In the thermolysine catalyzed reaction between benzyloxycarbonyl-L-aspartic acid and L-phenyl-alanine methyl ester, used in excess, at pH 7 the equilibrium is shifted to the right

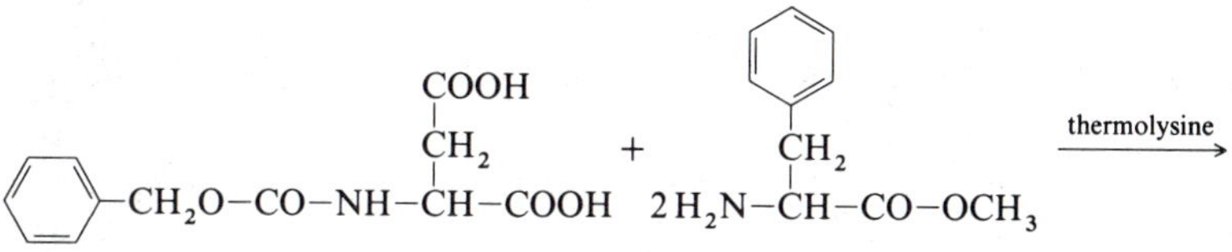

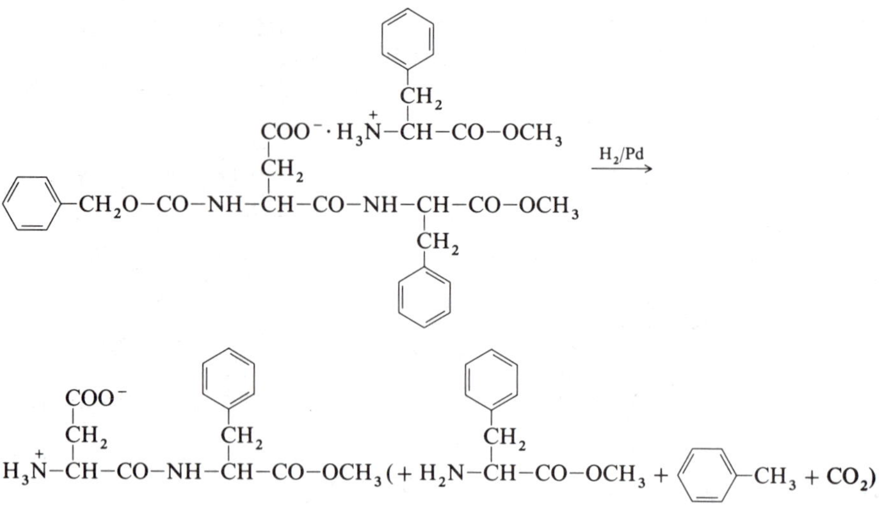

$$\text{CH}_2(\text{phenyl})$$

$$\text{COO}^- \cdot \overset{+}{\text{H}_3\text{N}} - \text{CH} - \text{CO} - \text{OCH}_3$$

$$\text{CH}_2$$

$$\text{(phenyl)} - \text{CH}_2\text{O} - \text{CO} - \text{NH} - \text{CH} - \text{CO} - \text{NH} - \text{CH} - \text{CO} - \text{OCH}_3 \xrightarrow{\text{H}_2/\text{Pd}}$$

$$\text{CH}_2(\text{phenyl})$$

$$\text{COO}^-$$

$$\text{CH}_2 \qquad \text{CH}_2(\text{phenyl})$$

$$\overset{+}{\text{H}_3\text{N}} - \text{CH} - \text{CO} - \text{NH} - \text{CH} - \text{CO} - \text{OCH}_3 (+ \text{H}_2\text{N} - \overset{\text{CH}_2(\text{phenyl})}{\text{CH}} - \text{CO} - \text{OCH}_3 + \text{(phenyl)} - \text{CH}_3 + \text{CO}_2)$$

because the product separates from the reaction mixture as an insoluble salt. Preparation of the target compound is concluded with the removal of the benzyloxycarbonyl group by hydrogenolysis. A more ambitious endeavor, synthesis of the nonapeptide amide oxytocin

Cys—Tyr—Ile—Gln—Asn—Cys—Pro—Leu—Gly—NH$_2$

required the use of several proteolytic enzymes: carboxypeptidase Y, chymotrypsin, chymopapain, trypsin and postproline-specific endopeptidase. Combination of enzymatic coupling with chemical methods of peptide bond formation turned out to be more practical. In general terms it might be stated, that the specificity of enzymes leads to great variations in coupling rates in reactions involving different amino acids. Hence, it seems to be likely that for the synthesis of various peptide bonds individual optimal conditions must be established and perhaps even the enzyme has to be judiciously selected. Furthermore, the hydrolytic potential of both the enzyme and the conditions applied must be carefully scrutinized: it would be rather unfortunate to form a new peptide bond at the expense of cleaving another. It appears at this time that enzyme catalyzed synthesis will play a major role in the preparation of certain selected target compounds but it is less likely that it will replace the methods of organic chemistry in the general practice of peptide synthesis.

## G. Non-conventional Methods of Peptide Bond Formation

The remarkable facile synthesis of proteins in the ribosomes of cells seems to act as a stimulus for organic chemists to invent novel alternatives in peptide synthesis. Two examples are mentioned here as illustrations of ingenuity although both methods lack practical application at this time.

Acid catalyzed rearrangement of O-aminoacyl salicylamides (Brenner et al. 1955)

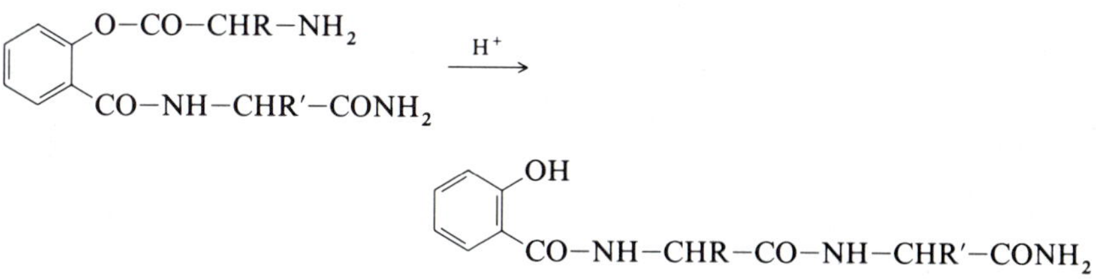

$$\begin{array}{c}\text{O—CO—CHR—NH}_2\\ \\ \text{CO—NH—CHR'—CONH}_2\end{array} \quad \xrightarrow{\text{H}^+}$$

$$\begin{array}{c}\text{OH}\\ \\ \text{CO—NH—CHR—CO—NH—CHR'—CONH}_2\end{array}$$

and the somewhat analogous reaction of acylaminoacyl-aminoacyl-hydrazines (Brenner and Hofer 1961)

$$\text{Y—NH—CHR—CO—NH—NH—CO—CHR'—NH}_2 \xrightarrow{\text{H}^+}$$

$$\text{Y—NH—CHR—CO—NH—CHR'—CO—NH—NH}_2$$

entail the *insertion* of an amino acid residue into a peptide bond. In the perhaps even more surprising four center condensation (4CC) method (Ugi 1980) a new amino acid is generated from an aldehyde:

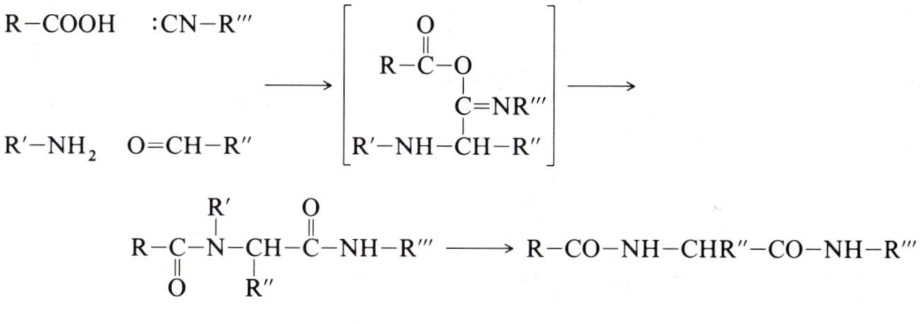

$$\begin{array}{ll}\text{R—COOH} & \text{:CN—R'''}\\ \\ \\ \text{R'—NH}_2 & \text{O=CH—R''}\end{array} \longrightarrow \left[\begin{array}{c}\text{O}\\ \parallel\\ \text{R—C—O}\\ |\\ \text{C=NR'''}\\ |\\ \text{R'—NH—CH—R''}\end{array}\right] \longrightarrow$$

$$\begin{array}{c}\text{R'} \quad \text{O}\\ |\quad\quad \parallel\\ \text{R—C—N—CH—C—NH—R'''}\\ \parallel \quad\; |\\ \text{O} \quad\; \text{R''}\end{array} \longrightarrow \text{R—CO—NH—CHR''—CO—NH—R'''}$$

With the utilization of chiral starting materials it became possible to achieve induced asymmetric synthesis and obtain peptides in enantiomerically pure form.

An equally interesting and perhaps more practical approach to coupling is the oxidation-reduction method (Mukaiyama et al. 1968)

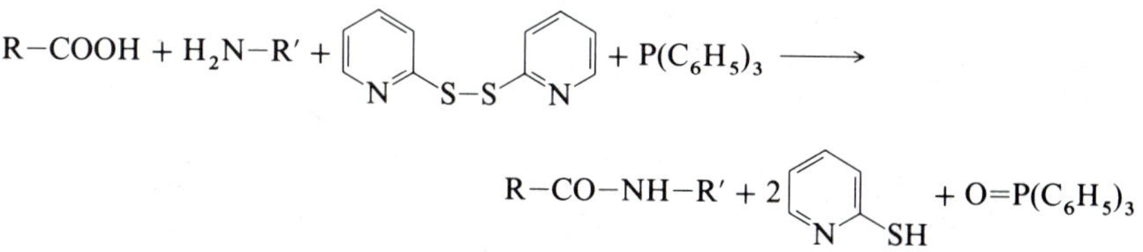

$$\text{R—COOH} + \text{H}_2\text{N—R'} + \underset{\text{N} \quad \text{S—S} \quad \text{N}}{\bigcirc\bigcirc} + \text{P(C}_6\text{H}_5)_3 \longrightarrow$$

$$\text{R—CO—NH—R'} + 2\,\underset{\text{N} \quad \text{SH}}{\bigcirc} + \text{O=P(C}_6\text{H}_5)_3$$

which probably involves an acylphosphonium intermediate:

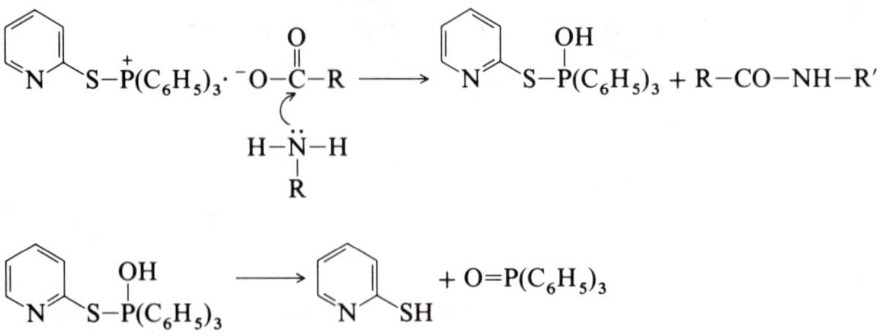

These procedures demonstrate that for inventive minds peptide bond formation is not a closed chapter.

# References

Anderson, G. W., Zimmerman, J. E., Callahan, F. M.: J. Amer. Chem. Soc. *85*, 3039 (1963)

Bates, A. J., Galpin, I. J., Hallett, A., Hudson, D., Kenner, G. W., Ramage, R.: Helv. Chim. Acta *58*, 688 (1975)

Belleau, D., Malek, G.: J. Amer. Chem. Soc. *90*, 1651 (1968)

Bodanszky, M.: Nature *175*, 685 (1955)

Boissonnas, R. A.: Helv. Chim. Acta. *34*, 874 (1951)

Brenner, M., Hofer, W.: Helv. Chim. Acta *44*, 1794, 1798 (1961)

Brenner, M., Zimmermann, J. P., Wehrmüller, J., Quitt, P., Photaki, I.: Helv. Chim. Acta *40*, 1497 (1957)

Castro, B., Dormoy, J. R., Evin G., Selve, C.: Tetrahedron lett. *1975*, 1219

Curtius, T.: Ber. dtsch. Chem. Ges. *35*, 3226 (1902)

Fischer, E.: Ber. dtsch. Chem. Ges. *36*, 2094 (1903)

Fruton, J. S. in Advances in Enzymology, vol. 53 Meister, A. ed. p. 239, New York, Wiley 1982

König, W., Geiger, R.: Chem. Ber. *103*, 788, 2024, 2034 (1970)

König, W., Geiger, R.: Chem. Ber. *186*, 3626 (1973)

Kovács, J., Kisfaludy, L., Ceprini, M. Q.: J. Amer. Chem. Soc. *89*, 183 (1967)

Kupriszewski, G.: Rocz. Chem. *35*, 595 (1961); Chem. Abstr. *55*, 27121i (1961)

Leuchs, H.: Ber. dtsch. Chem. Ges. *69*, 857 (1906)

Mukaiyama, T., Ueki, M., Maruyama, H., Matsueda, R.: J. Amer. Chem. Soc. *90*, 4490 (1968); ibid. *91*, 1554 (1969)

Pless, J., Boissonnas, R. A.: Helv. Chim. Acta *46*, 1609 (1963)

Sealock, R. W., Laskowski, M., Jr.: Biochemistry *8*, 3703 (1969); cf. also Laskowski, M. in Semisynthetic Peptides and Proteins, Offord, R. E., DiBello, C., eds. p. 263, London, Acad. Press 1978

Schwyzer, R., Iselin, B., Feurer, M.: Helv. Chim. Acta *38*, 69 (1955); Schwyzer, R., Feurer, M., Iselin, B., Kagi, H. ibid. *38*, 80 (1955); Schwyzer, R., Feurer, M., Iselin, B.: ibid. *38*, 1067 (1955)

Sheehan, J. C., Hess, G. P.: J. Amer. Chem. Soc. *77*, 1067 (1955)

Staab, H. A.: Liebigs Ann. Chem. *609*, 75 (1957)
Ugi, I. in The Peptides, vol. 2 (Gross, E., Meienhofer, J., eds.) p. 365, New York, Academic Press 1986
Vaughan, J. R., Jr.: J. Amer. Chem. Soc. *73*, 3547 (1951)
Vaughan, J. R., Osato, R. L.: J. Amer. Chem. Soc. *73*, 5553 (1951)
Wieland, T., Bernhard, H.: Liebigs Ann. Chem. *572*, 190 (1951)

# Additional Sources

Bodanszky, M.: Active Esters in Peptide Synthesis, in The Peptides, vol. 1, Gross, E., Meienhofer, J. eds., pp. 105–196, New York, Academic Press 1979
Meienhofer, J.: The Azide Method in Peptide Synthesis, in The Peptides, vol. 1, Gross, E., Meienhofer, J. eds., pp. 197–239, New York Academic Press 1979
Meienhofer, J.: The Mixed Anhydride Method of Peptide Synthesis, in The Peptides vol. 1, Gross, E., Meienhofer, J. eds., pp. 263–314, New York, Academic Press 1979

# VI. Protection of Functional Groups

## A. Blocking of α-Amino Groups

While it is necessary to mask the amino function of an amino acid or peptide during activation of the carboxyl group which participates in the subsequent coupling, it is equally important to select protecting groups that can readily be removed. Otherwise the integrity of the already formed peptide bonds would be endangered. If the peptide obtained in a coupling reaction has to be lengthened at its N-terminus, then an additional requirement limits the choice of the amine-blocking group: it must be removable under conditions which leave the masking of the carboxyl group and the protection of side chain functions intact. Thus, blocking of the α-amino function is generally *transient* while that of all other functions mostly *semipermanent*.

$$Y-NH-CHR'-CO-X + H_2N-CHR-COOY' \xrightarrow[(-HX)]{coupling}$$

$$Y-NH-CHR'-CO-NH-CHR-COOY' \xrightarrow{partial\ deprotection}$$

$$H_2N-CHR'-CO-NH-CHR-COOY' \xrightarrow{acylation\ with\ Y-NH-CHR''-CO-X}$$

$$Y-NH-CHR''-CO-NH-CHR'-CO-NH-CHR-COOY' \xrightarrow{complete\ deblocking}$$

$$H_2N-CHR''-CO-NH-CHR'-CO-NH-CHR-COOH$$

X = activating group
Y = transient amine-protecting group
Y' = semipermanent carboxyl-protecting group

Several simple amine-protecting groups derived from carboxylic acids and commonly used in organic synthesis are obviously not suitable in peptide synthesis. For instance, acetylation or benzoylation of amino groups is impractical, because the vigorous hydrolysis needed for deacylation cleaves peptide bonds as well. The formyl and trifluoroacetyl groups were found to be useful but even these are employed mainly for the masking of side chain amine groups and will be discussed in that connection. On the other hand derivatives of *carbonic acid* gradually reached a position of monopoly. Their effectiveness is based on the lability of *carbamoic acids* which spontaneously lose carbon dioxide:

$$R-NH-COOH \longrightarrow R-NH_2 + CO_2$$

This opened up the attractive possibility of incorporating the amino group into an ester of carbamoic acid. In the resulting *urethane* cleavage of the ester and the ensuing decarboxylation regenerate the free amine:

$$R-NH-CO-OR' \xrightarrow[(-R'OH)]{HOH} R-NH-COOH \longrightarrow R-NH_2 + CO_2$$

Yet, the earliest attempts in this direction (Emil Fischer 1902) failed. Urethanes, in which polarization of the carbonyl group is counterbalanced by the availability of electrons from the neighboring oxygen *and* nitrogen atoms, are markedly resistant to nucleophiles, including the hydroxyl ion. Hence, during the treatment of ethyloxycarbonyl-glycyl-glycine ethyl ester with alkali only the ester group at the C-terminal carboxyl was saponified, while the urethane grouping lost its ethyloxy group by intramolecular displacement resulting in cyclization to a hydantoin which, in turn, opened up to yield a urea derivative:

$$H_5C_2O-CO-NH-CH_2-CO-NH-CH_2-CO-OC_2H_5 \xrightarrow{OH^-}$$

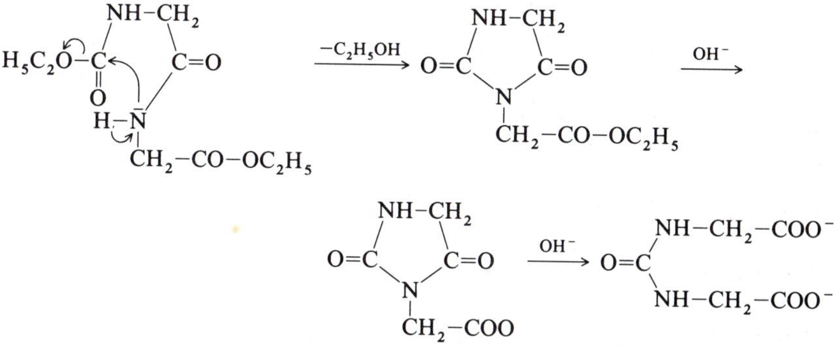

It was, therefore, an invention of major significance when in 1932 Bergmann and Zervas proposed replacement of the ethyloxycarbonyl group by the benzyloxycarbonyl (or "carbobenzoxy" or "carbobenzyloxy", Cbz or Z) group. The latter smoothly undergoes homolytic fission on catalytic hydrogenation. The reaction is concluded with spontaneous decarboxylation:

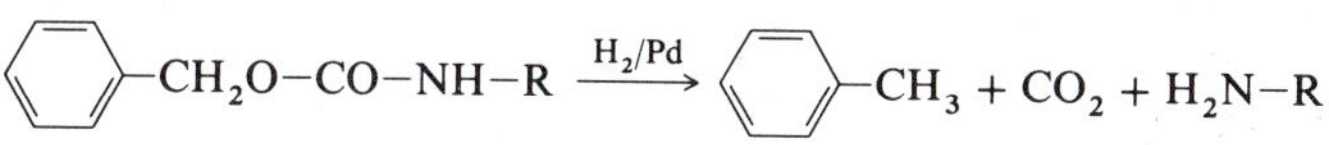

Hydrogenolysis of benzyl esters is catalyzed by platinum metals, particularly by palladium. The latter is applied as palladium black or palladium on charcoal or palladium on barium sulfate. In addition to the most important feature of the method, stability of the peptide bond under the conditions of the deblocking procedure, the relatively harmless nature of the by-products, toluene and carbon dioxide, both removed without difficulty, rendered the new protecting group extremely attractive. Instead of hydrogen gas, hydrogen donors such as hydrazine, formic acid or ammonium formate can also be applied, but these are not fully inert toward peptides. The more innocuous cyclohexene requires elevated temperatures for rapid hydrogenolysis while the very efficient cyclohexadiene is fairly expensive. Thus the classical process of hydrogenation in the presence of palladium on charcoal remains the method of choice in most laboratories. Other reductive methods, for instance reduction with sodim in liquid ammonia, are less frequently practiced.

The recognition (Ben Ishai and Berger 1951) that the benzyloxycarbonyl group is cleaved by *acidolysis,* particularly by HBr in acetic acid, had a major impact on the development of protecting groups. In the heterolytic fission of the carbon-oxygen bond the *benzyl cation* plays an important role by facilitating the decomposition of the protonated itermediate:

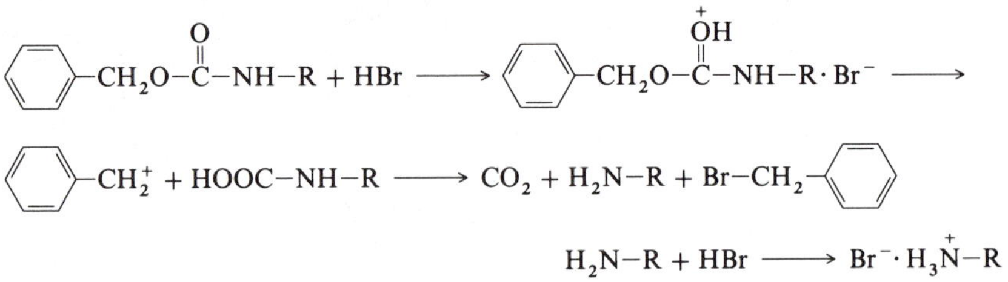

Neither the lachrimatory effect of benzyl bromide, the by-product of the reaction, nor its ability to alkylate various amino acid side chains should escape attention.

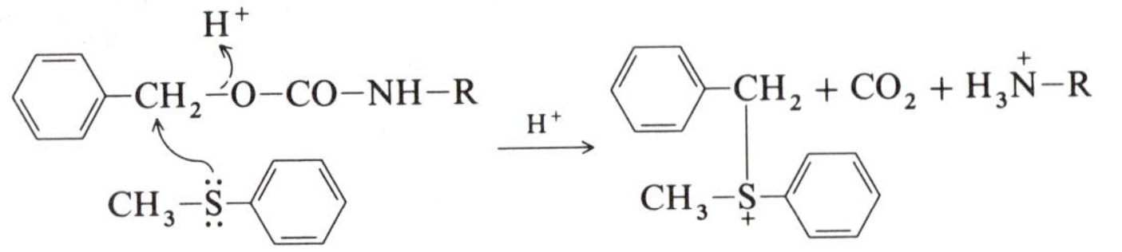

The classical reagent, HBr in acetic acid, can be replaced by HBr in trifluoroacetic acid, HBr in liquid $SO_2$ or by liquid $SO_2$ itself. Of course the benzyloxycarbonyl group is rapidly cleaved by liquid HF or by a solution of trifluoromethanesulfonic acid in trifluoroacetic acid. It should be mentioned that it is removed by the much less aggressive trifluoroacetic acid as well. At elevated temperatures the reaction proceeds fairly rapidly, at room temperature it requires several days for completion. Addition of thioanisole (Yajima et al 1978) greatly accelerates the process and similar results could be achieved with 4-methylthiophenol,

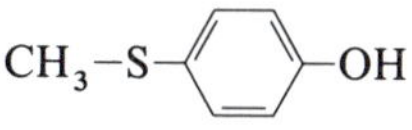

$$CH_3-S-\langle\ \rangle-OH$$

a less volatile scavenger-catalyst.

Soon it became obvious that acid sensitivity of the benzyloxycarbonyl group can be increased by appropriate modifications of the benzyl group. Electron-releasing substituents give rise to more stable cations, hence the p-methoxybenzyloxycarbonyl group

$$CH_3-O-\langle\ \rangle-CH_2O-CO-NH-R$$

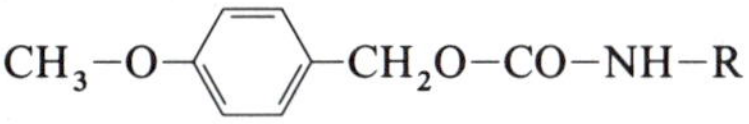

is cleaved by weaker acids, such as dilute solutions of HCl in acetic acid and also by trifluoroacetic acid. Similar acid-sensitivity is afforded by the replacement of the benzyl group with the tert.butyl group, source of a much more stable cation:

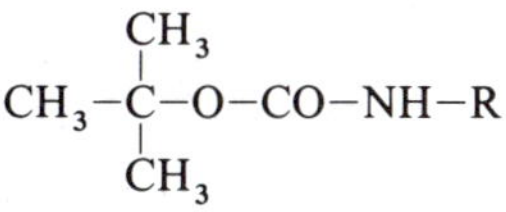

$$\begin{array}{c} CH_3 \\ | \\ CH_3-C-O-CO-NH-R \\ | \\ CH_3 \end{array}$$

Further modifications in the same direction culminated in the highly acid labile biphenylylisopropyloxycarbonyl (BPoc) group (Sieber and Iselin 1968)

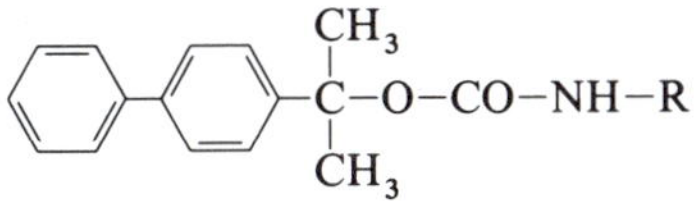

$$\langle\ \rangle-\langle\ \rangle-\begin{array}{c} CH_3 \\ | \\ C-O-CO-NH-R \\ | \\ CH_3 \end{array}$$

which served well in several demanding syntheses.

Sensitivity to acids is not restricted to urethane-type blocking groups. Thus, the triphenylmethyl (trityl, Trt) group is cleaved by acetic acid and also by tetrazole in trifluoroethanol

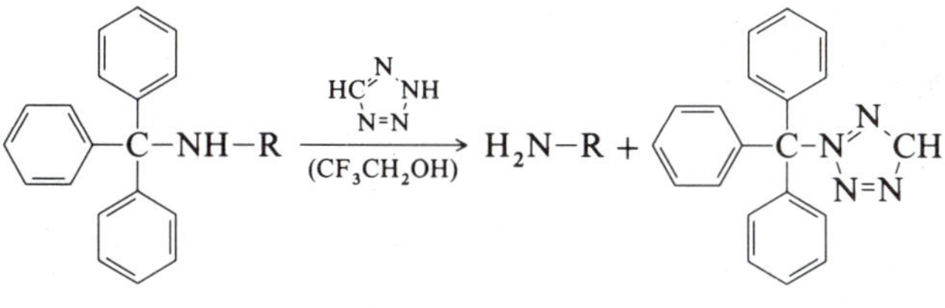

while the *o*-nitrobenzenesulfenyl (Nps) group by dilute solutions of HCl in organic solvents:

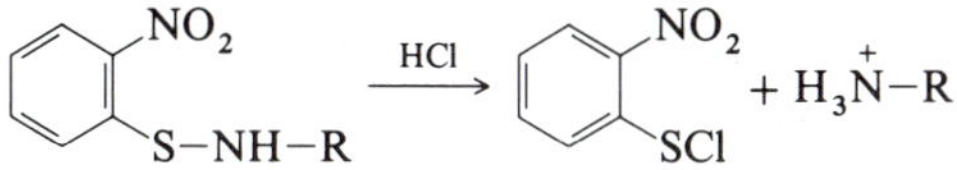

Amines blocked by the Nps group can be regenerated by thiolysis as well, for instance with 2-mercaptopyridine:

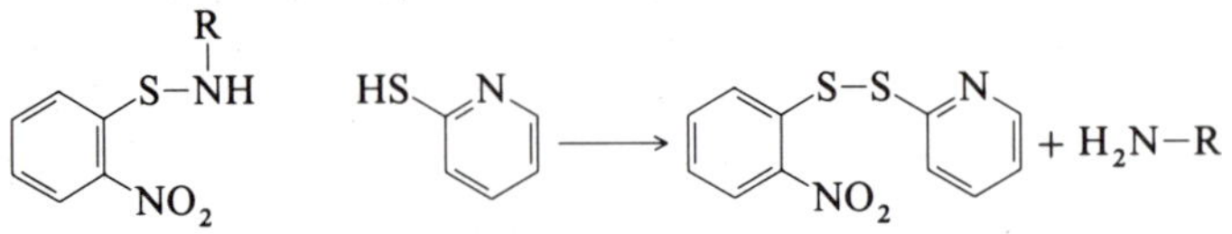

A recently proposed series of reagents, derivatives of trimethylsilane

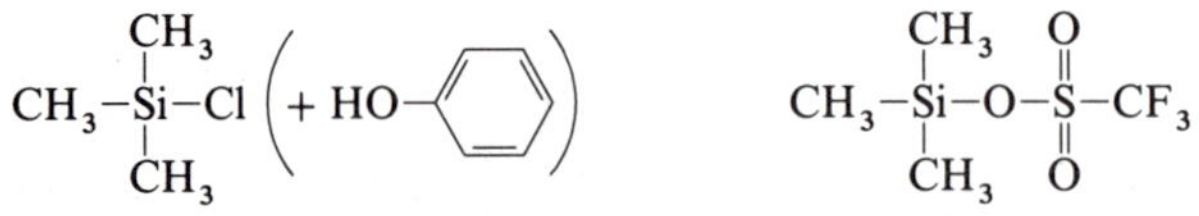

offer potentially useful alternatives for the removal of acid labile blocking groups, such as the tert.butyloxycarbonyl (Boc) group.

In order to design syntheses of complex peptides containing several functional side chains, it is necessary to have a variety of methods of protection and deblocking at disposal. Therefore, it was a major step forward when in addition to reduction and acidolysis deprotection with weak *bases,* under mild conditions, became a practical possibility. The 9-fluorenylmethyloxycarbonyl (Fmoc) group (Carpino and Han 1970) is removed from the amino group by proton abstraction with secondary amines. A carbamoic acid is generated, that, in turn, loses carbon dioxide and affords the free amine:

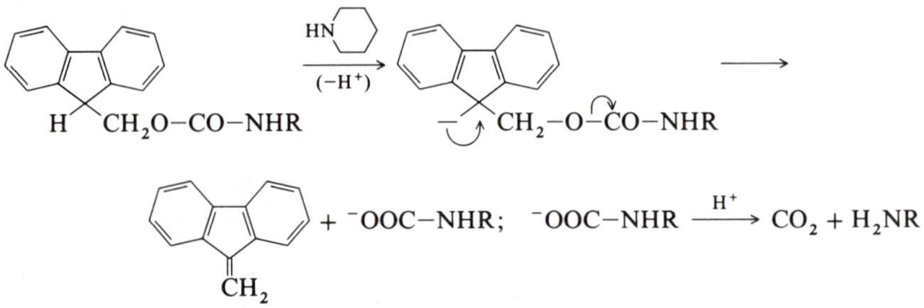

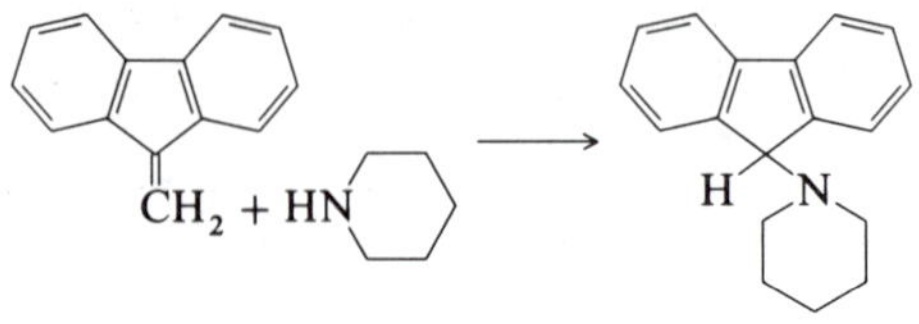

The fact that the amino groups is indeed free and not protonated as in deprotection by acidolysis, is a distinct advantage of the method. Piperidine is the recommended base although in some cases better results can be obtained with diethylamine. The by-product, dibenzofulvene, combines with the secondary amine to yield a stable teriary base which, however, usually does not interfere with the reaction or with isolation of the product.

Development of amine-protecting groups is not a closed chapter. That novel blocking groups, sensitive to specific reagents, can be designed was shown by Carpino and his associates (1978) who proposed the trimethylsilylethyloxy-carbonyl group which is selectively cleaved by the fluoride anion applied in the form of tetraalkylammonium fluorides:

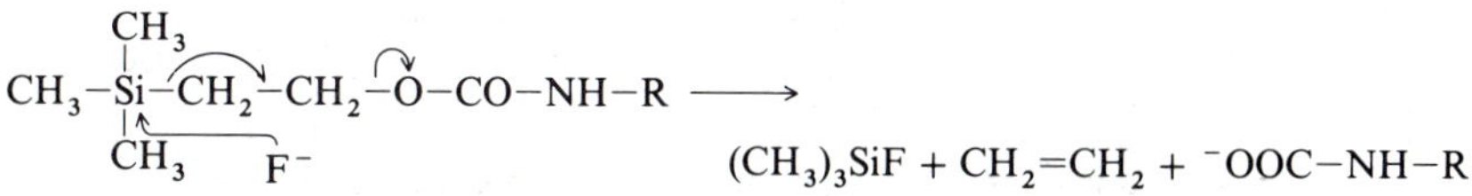

$$CH_3-\underset{\underset{CH_3}{|}}{\overset{\overset{CH_3}{|}}{Si}}-CH_2-CH_2-O-CO-NH-R \longrightarrow (CH_3)_3SiF + CH_2{=}CH_2 + {}^-OOC-NH-R$$

A similarly distinct selectivity is characteristic for the di-*p*-nitrophenyl-ethyl-oxycarbonyl group

$$O_2N{-}\langle\text{Ar}\rangle{-}\overset{|}{CH}-CH_2O-CO-NH-R,\ \ \overset{|}{\underset{O_2N{-}\langle\text{Ar}\rangle}{}}$$

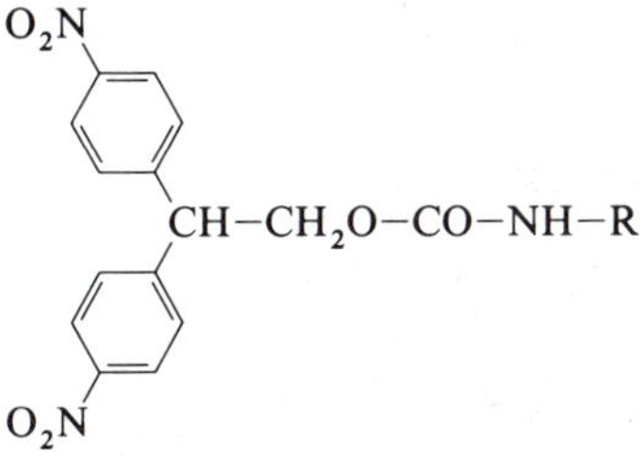

which is stable toward acids and alkali and also secondary amines, such as dialkylamines or piperidine, but is smoothly cleaved by the strong cyclic bases

even if these are buffered with acetic acid.

Before concluding the discussion of protection of the amino group we have to mention the methods used for the preparation of N-blocked amino acid derivatives. The classical procedure, still applied for the introduction of the benzyloxycarbonyl group, is acylation with the appropriate alkyl chlorocar-bonate[1]. These, in turn, are obtained through the reaction of phosgene with the alcohol:

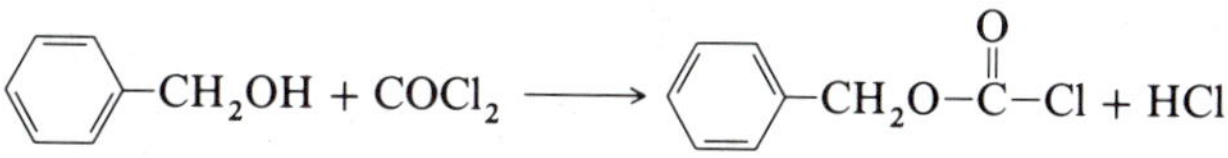

$$\langle\text{Ph}\rangle{-}CH_2OH + COCl_2 \longrightarrow \langle\text{Ph}\rangle{-}CH_2O-\overset{\overset{O}{\|}}{C}-Cl + HCl$$

After removal of the HCl generated in the reaction (preferably with a stream of nitrogen) the reagent is applied without purification. Distillation of benzyl

---

[1] The more commonly used expression "chloroformate" while formally correct, does not point to carbonic acid. Yet, protection in the form of urethanes is based on carbonic acid chemistry.

chlorocarbonate is possible and sometimes desirable, but must be carried out in high vacuum, because at higher temperature and also on storage the reagent disproportionates to phosgene and dibenzyl carbonate. The chlorocarbonate is added to the aqueous solution of the amino acid in the presence of alkali

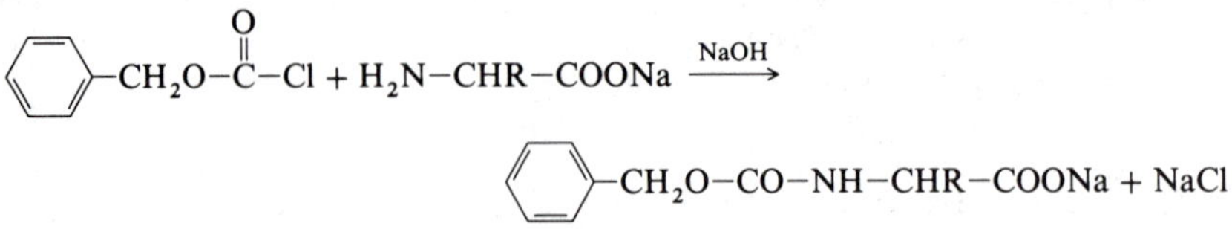

$$\text{Ph–CH}_2\text{O–CO–Cl} + \text{H}_2\text{N–CHR–COONa} \xrightarrow{\text{NaOH}}$$

$$\text{Ph–CH}_2\text{O–CO–NH–CHR–COONa} + \text{NaCl}$$

and the blocked amino acids are obtained by acidification followed by extraction or crystallization. This simple approach is not practical in the case of the tert.butyloxycarbonyl (Boc) group because tert.butyl chlorocarbonate is not stable enough for isolation and storage. Hence the carbonic acid half ester-half azide was prepared via the phenyl ester and hydrazide

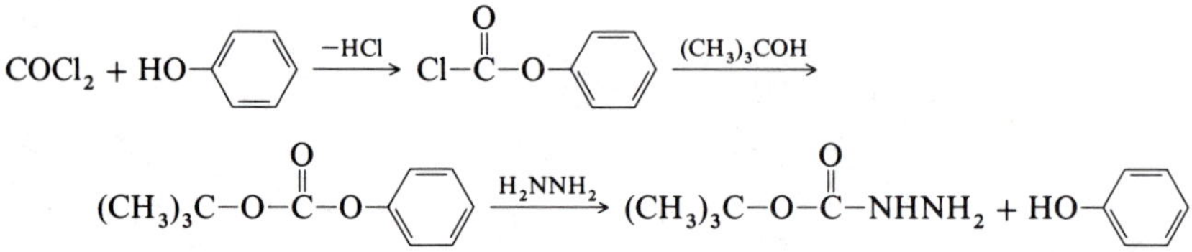

$$\text{COCl}_2 + \text{HO–Ph} \xrightarrow{-\text{HCl}} \text{Cl–CO–O–Ph} \xrightarrow{(\text{CH}_3)_3\text{COH}}$$

$$(\text{CH}_3)_3\text{C–O–CO–O–Ph} \xrightarrow{\text{H}_2\text{NNH}_2} (\text{CH}_3)_3\text{C–O–CO–NHNH}_2 + \text{HO–Ph}$$

and the latter converted to the azide (tert.butyl azidoformate), an efficient acylating agent:

$$(\text{CH}_3)_3\text{C–O–CO–NHNH}_2 \xrightarrow{\text{HONO}} (\text{CH}_3)_3\text{C–O–CO–N}{=}\overset{+}{\text{N}}{=}\overset{-}{\text{N}}$$

$$(\text{CH}_3)_3\text{C–O–CO–N}{=}\overset{+}{\text{N}}{=}\overset{-}{\text{N}} + \text{H}_2\text{N–R} \longrightarrow (\text{CH}_3)_3\text{C–O–CO–NH–R} + \text{HN}_3$$

The use of the azide, however, created serious problems. Occasionally explosions were reported and the hydrazoic acid escaping during acylation is fairly toxic. Therefore, Boc derivatives of amino acids are often prepared with the help of reactive esters. From the numerous reagents proposed for this purpose only a few are shown here

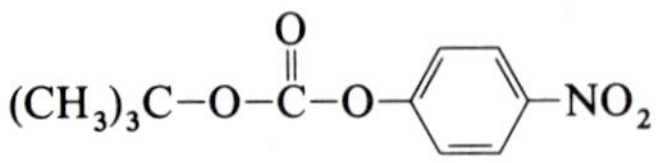

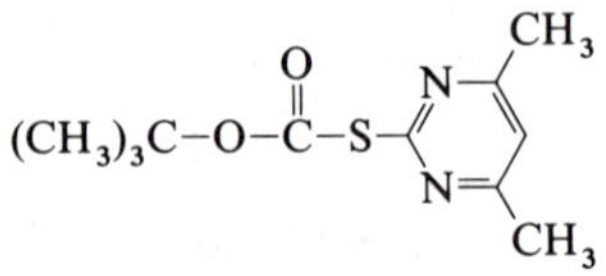

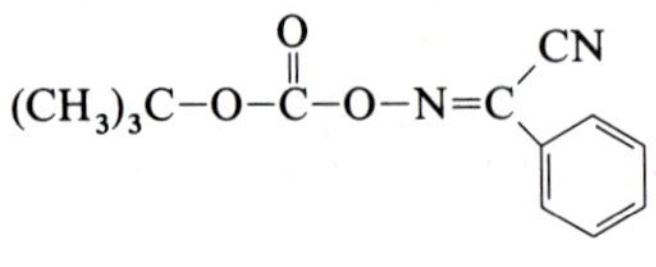

$$(\text{CH}_3)_3\text{C–O–CO–O–C}_6\text{H}_4\text{–NO}_2$$

$$(\text{CH}_3)_3\text{C–O–CO–S–}\left(\text{pyrimidine, CH}_3, \text{CH}_3\right)$$

$$(\text{CH}_3)_3\text{C–O–CO–O–N}{=}\text{C(CN)(C}_6\text{H}_5)$$

More recently tert.butyl pyrocarbonate, usually described as tert.butyl carbonic anhydride, gained considerable popularity, mainly because the products formed during introduction of the Boc group, are harmless and easily removed:

$$(CH_3)_3C-O-\overset{\overset{\displaystyle O}{\|}}{C}-O-\overset{\overset{\displaystyle O}{\|}}{C}-O-C(CH_3)_3 + H_2NR \longrightarrow$$

$$(CH_3)_3C-O-CO-NHR + (CH_3)_3COH + CO_2$$

The methods discussed here are applicable for the incorporation of other urethane-type blocking groups as well. In some cases the old chloride method prevails. For instance 9-fluorenylmethyl chlorocarbonate, a stable crystalline solid, allows smooth introduction of the Fmoc group.

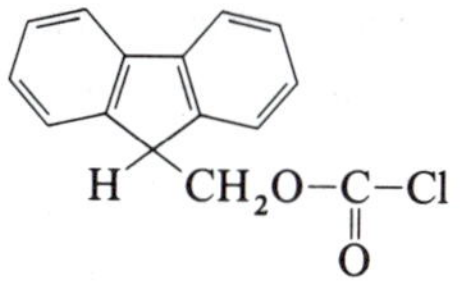

## B. Blocking of Side Chain Amino Groups

In order to prevent branching in the growing peptide chain one must block the amino group in the side chain of lysine (and ornithine) residues. The blocking group applied for this purpose is expected to remain in place during the chain building process, hence it must resist the reagents applied for the deprotection of α-amino groups and should be cleaved only at the completion of the synthetic scheme. Obviously an amine-protecting group different from the one used for the transient blocking of the α-amino groups is needed. A classical combination applied in the synthesis of larger peptides is the protection of side chain amino groups with the tert.butyloxycarbonyl (Boc) group and the α-amino functions with the benzyloxycarbonyl (Z) group. The coupling steps are followed by catalytic hydrogenation which unmasks the α-amino group and allows its acylation in the next chain-lengthening operation. The Boc group, unaffected during hydrogenolysis, is then cleaved at the conclusion of the synthesis, most frequently with trifluoroacetic acid:

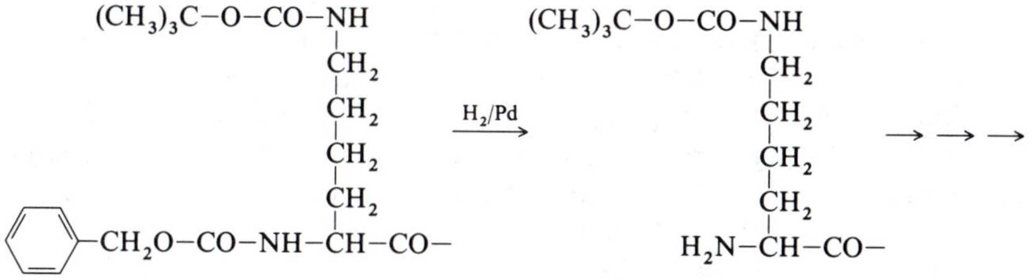

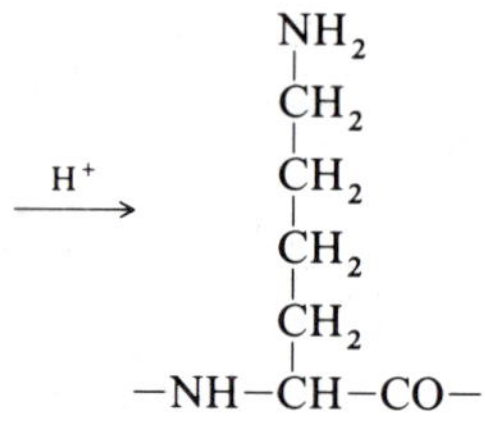

$$
\xrightarrow{\text{H}^+}
\begin{array}{l}
\text{NH}_2 \\
|\phantom{x} \\
\text{CH}_2 \\
|\phantom{x} \\
\text{CH}_2 \\
|\phantom{x} \\
\text{CH}_2 \\
|\phantom{x} \\
\text{CH}_2 \\
|\phantom{x} \\
-\text{NH}-\text{CH}-\text{CO}-
\end{array}
$$

In a sense this is a perfect example of *orthogonal* protection schemes in which two different chemical reactions are applied, one for the removal of the transient and the other for the cleavage of the semipermanent blocking groups. Unfortunately this combination can not be readily utilized in the synthesis of peptides with sulfur containing amino acid residues: poisoning of the catalyst impedes the process. It is possible to employ countermeasures, such as addition of base or the use of liquid ammonia as solvent, when methionine residues are the source of difficulties, but cysteine residues present a less readily surmountable obstacle.

Application of these groups in the opposite sense, that is the Z group for side chain protection of lysine residues and the Boc group for the transient masking of the α-amino group, is more generally applicable.

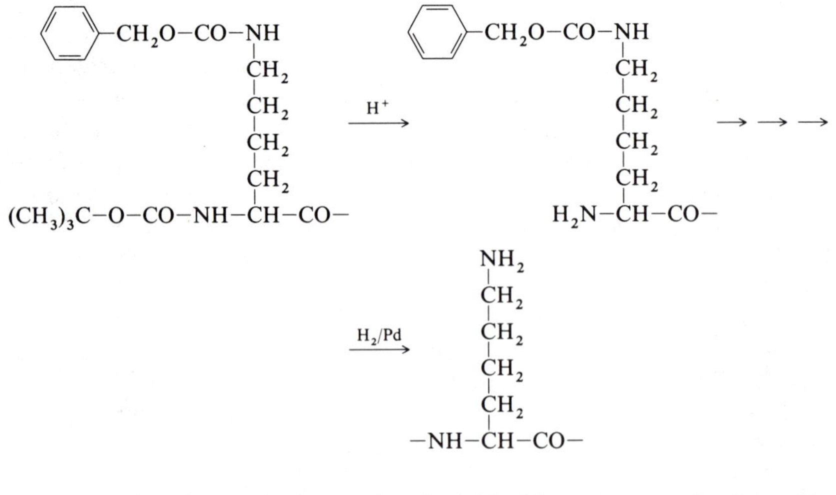

This is a non-orthogonal scheme, since both blocking groups can be cleaved by acids, one under mild, the other under more drastic conditions. Accordingly the combination of Z and Boc groups in this manner is imperfect: a small but not negligible loss of Z-group from lysine side chains occurs at the repeated cleavage of Boc groups. This imperfection appears to be minor, but it is the source of disturbing by-products. A remedy for this problem is to render the Z group more acid-resistant by destabilizing the benzyl cation with electron-withdrawing substituents. The modified Z-groups,

$$O_2N-\!\!\!\bigcirc\!\!\!-CH_2O-CO-NHR \qquad Cl-\!\!\!\bigcirc\!\!\!-CH_2O-CO-NHR$$

$$\bigcirc\!\!\!-CH_2O-CO-NHR \qquad Cl-\!\!\!\bigcirc\!\!\!-CH_2O-CO-NHR$$
$$\;\;Cl \qquad\qquad\qquad\qquad\qquad\;\;Cl$$

however, require very strong acids, such as hydrogen fluoride or trifluoromethanesulfonic acid for their removal in the final deprotection. These strong acids, then, by protonating even very weakly basic functions in the already assembled peptide, initiate various side reactions discussed in Chapter VII.

An alternative way to enhance selectivity in the removal of two different blocking groups is to increase the sensitivity of one of them toward acids. This thought led to the development of the already mentioned biphenylylisopropyloxycarbonyl (Bpoc) group and to the application of the highly acid sensitive triphenylmethyl (trityl) group.

From this discussion it follows, that orthogonal combinations are superior to schemes based on differences in sensitivity toward the same reagent. A particularly promising and already well accepted orthogonal combination is the use of the base labile 9-fluorenylmethyloxycarbonyl (Fmoc) group for the transient protection of $\alpha$-amino functions and of the tert.butyloxycarbonyl group for the masking of the side chain amino group of lysine residues. The carboxyls, including the carboxyl group at the C-terminus, are blocked in the form of tert.butyl esters, the hydroxyl groups as tert.butyl ethers.

At this point a further and quite interesting possibility must be mentioned. The side chain amino group of lysine can be provided with a blocking group which is stable under the conditions used for the removal of acid and base labile masking groups, but is cleaved by enzyme-catalyzed hydrolysis. Selective $N^\varepsilon$-acetyllysine hydrolases are not readily available, but the well studied penicillinamidohydrolase is an efficient catalyst in the hydrolysis of $N^\varepsilon$-phenylacetyllysine. Similarly, $N^\varepsilon$-pyroglutamyl-lysine is hydrolyzed in the presence of pyrrolidonecarboxylpeptidase under mild conditions. These acyl groups can be kept intact throughout the synthesis and removed in the concluding step.

$$\begin{array}{c} CH_2 \\ O=C \;\;\;\;\; CH_2 \\ \backslash\;\;\;\;/ \\ NH-CH \end{array}$$

$$\bigcirc\!\!\!-CH_2-CO-NH \qquad\qquad CO-NH$$
$$\qquad\qquad\quad |\qquad\qquad\qquad\qquad\quad |$$
$$\qquad\qquad\quad CH_2\qquad\qquad\qquad\qquad CH_2$$
$$\qquad\qquad\quad CH_2\qquad\qquad\qquad\qquad CH_2$$
$$\qquad\qquad\quad CH_2\qquad\qquad\qquad\qquad CH_2$$
$$\qquad\qquad\quad CH_2\qquad\qquad\qquad\qquad CH_2$$
$$\qquad H_2N-CH-COOH \qquad\qquad H_2N-CH-COOH$$

Regioselective introduction of protecting groups in lysine poses no major problem. The copper(II) complex of the amino acid is acylated and the $N^\varepsilon$-acyl-derivative secured through decomposition of the complex, for instance with $H_2S$:

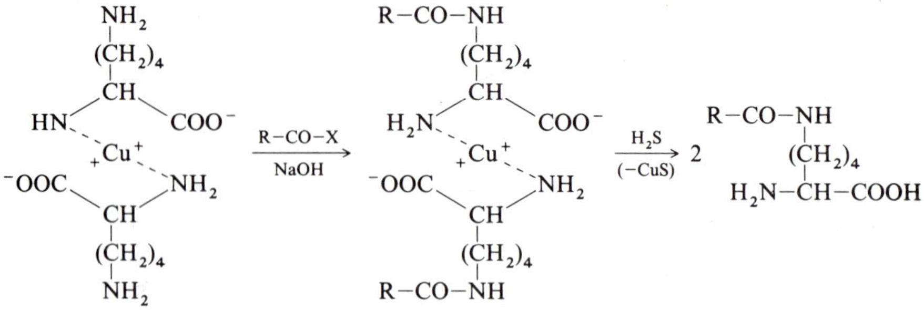

# C. Protection of the Carboxyl Group

The need to mask free carboxyl groups arises in two different conexts. Transient carboxyl protection of the amine-component is necessary in coupling steps in which the product will serve as carboxyl-component in the subsequent coupling of segments. Semipermanent blocking has to be used if the carboxyl belongs to the C-terminal residue of a target molecule: the masking group is removed only in the concluding step "final" deprotection. The same is true, of course for the side chain carboxyls of aspartyl and glutamyl residues, which have to remain blocked throughout the construction of the chain.

The general approach for carboxyl protection is esterification. The simplest solution, the use of methyl or ethyl esters, is suitable for semipermanent blocking, although the commonly applied process of unmasking, alkaline hydrolysis, is far from unequivocal. It is accompanied by racemization, partial hydrolysis of carboxamide groups in the side chain of asparagine and glutamine residues and by several other side reactions which are initiated by proton abstraction (Cf. Chapter VII). Nevertheless, perhaps because of the attractively simple esterification of amino acids

$$H_2N-CHR-COOH + CH_3OH \xrightarrow[\text{(HCl)}]{\bullet\, H^+} Cl^-\cdot H_3\overset{+}{N}-CHR-CO-OCH_3 + H_2O$$

this method of carboxyl protection remains in the practice of peptide synthesis. Further simplifaction of the acid catalyzed esterification can be achieved by replacing the introduction of gaseous HCl with the addition of thionyl chloride to the alcohol and even more conveniently through the use of p-toluenesulfonic acid as catalyst. The resulting amino acid ester p-toluenesulfonate salts are more tractable than the often hygroscopic hydrochlorides.

Saponification with alkali generally used for the removal of methyl and ethyl esters is avoided if instead of simple alkyl esters rather benzyl esters or tert.butyl esters are applied for the protection of carboxyl groups. In addition to acid catalyzed esterification benzyl esters can also be secured by the reaction of cesium salts of (blocked) amino acids with benzyl chloride:

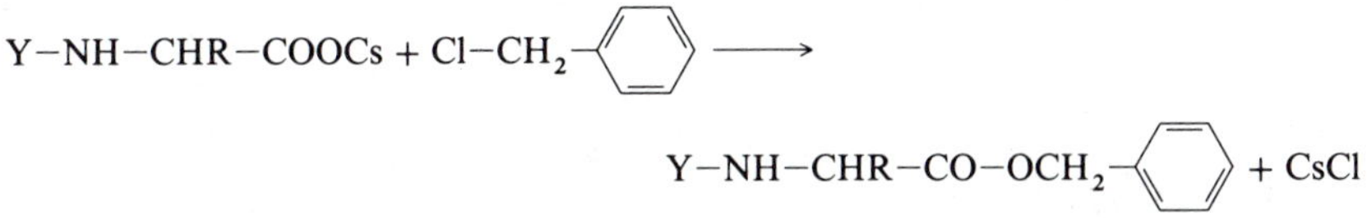

$$Y-NH-CHR-COOCs + Cl-CH_2-\langle\bigcirc\rangle \longrightarrow$$

$$Y-NH-CHR-CO-OCH_2-\langle\bigcirc\rangle + CsCl$$

Addition of the carboxyl group to isobutene, in the presence of strong acids, is used for the preparation of tert.butyl esters:

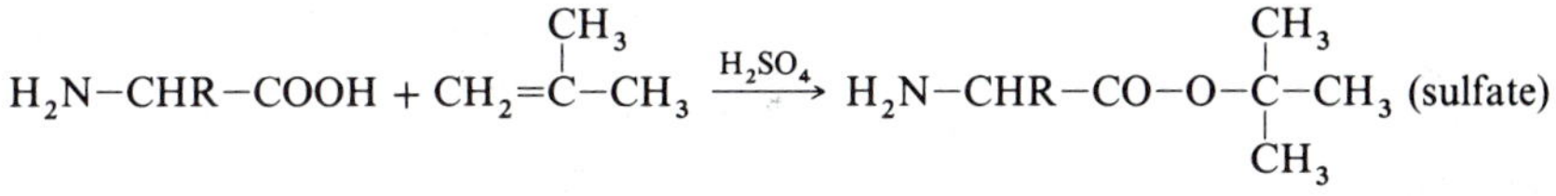

$$H_2N-CHR-COOH + CH_2=\overset{\overset{\displaystyle CH_3}{|}}{C}-CH_3 \xrightarrow{H_2SO_4} H_2N-CHR-CO-O-\overset{\overset{\displaystyle CH_3}{|}}{\underset{\underset{\displaystyle CH_3}{|}}{C}}-CH_3 \text{ (sulfate)}$$

The salts of tert-butyl esters thus obtained can be converted to the free amines which are stable and can be stored while free methyl, ethyl and benzyl esters undergo self-condensation to diketopiperazines:

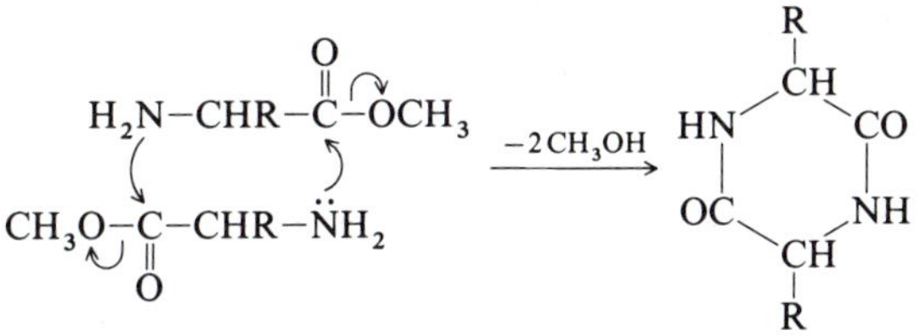

$$\begin{array}{c} H_2N-CHR-\overset{\overset{\displaystyle O}{||}}{C}-OCH_3 \\ \\ CH_3O-\overset{\underset{\displaystyle O}{||}}{C}-CHR-NH_2 \end{array} \xrightarrow{-2CH_3OH} \begin{array}{c} R \\ | \\ CH \\ HN \diagup \diagdown CO \\ | \quad\quad | \\ OC \diagdown \diagup NH \\ CH \\ | \\ R \end{array}$$

Benzyl esters are cleaved by catalytic hydrogenation and also by acidolysis albeit only under fairly drastic conditions. Electron-withdrawing substituents, such as the nitro group in p-nitrobenzyl esters, destabilize the benzyl cation and render the benzyl group even more resistant to acids. In contrast, the action of moderately strong acids, like dilute solutions of HCl in acetic acid or neat trifluoroacetic acid is sufficient for the acidolytic cleavage of p-methoxybenzyl esters and, of course, of tert.butyl esters.

So far only a few groups are known which are removable from the blocked carboxyl under mild alkaline conditions. Thus, phenyl esters are cleaved with alkaline hydrogen peroxide

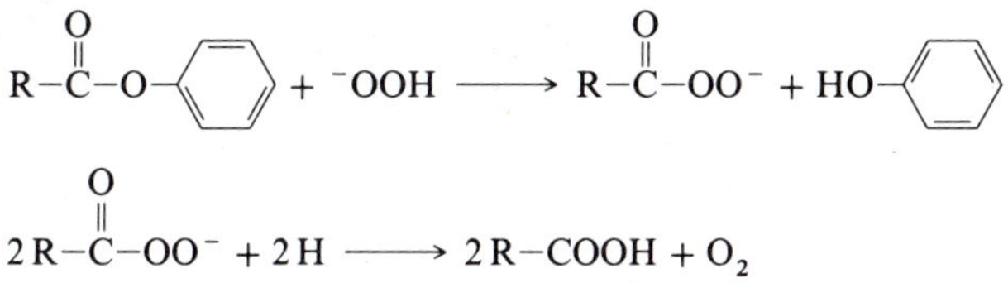

$$R-\overset{\overset{\displaystyle O}{||}}{C}-O-\langle\bigcirc\rangle + \text{ }^-OOH \longrightarrow R-\overset{\overset{\displaystyle O}{||}}{C}-OO^- + HO-\langle\bigcirc\rangle$$

$$2R-\overset{\overset{\displaystyle O}{||}}{C}-OO^- + 2H \longrightarrow 2R-COOH + O_2$$

and this method already found application in demanding syntheses. Yet, un-masking of the blocked carboxyl group is not a closed chapter. An interesting pathway involving intramolecular catalysis was designed (Barton et al. 1973) for the removal of peptides from polymeric supports (cf. Chapter X) to which they were anchored in form of (substituted) benzyl esters:

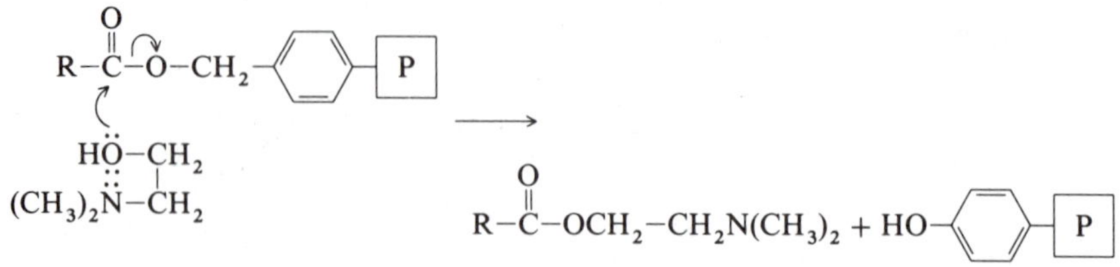

The attack by the nucleophile, dimethylaminoethanol, is followed by hydrolysis of the dimethylaminoethyl ester with water under relatively mild conditions: general base catalysis is provided by the tertiary amino group in the ester portion of the molecule:

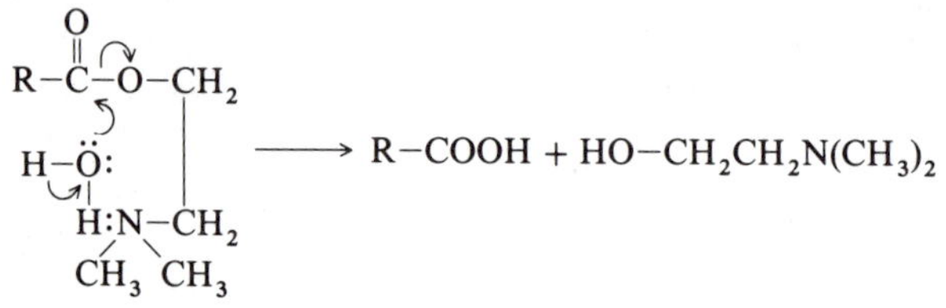

The idea of catalysis is even more apparent in the so far only sporadically practiced procedures in which proteolytic enzymes, such as trypsin, chy-motrypsin or carboxypeptidases are used for acceleration of the hydrolysis of alkyl esters under neutral conditions.

## D. Protection of Hydroxyl Groups

The secondary hydroxyl in threonine is mostly unaffected during coupling because the nucleophilic character of the hydroxyl group is less pronounced than that of the amino group and its reactivity is further decreased by steric hindrance from the nearby methyl group. Therefore blocking of the threonine side chain is not mandatory, but generally it is still provided with a masking group if the acylating agent is used in large excess, as in solid phase peptide synthesis (cf. Chapter X). The situation is somewhat different with the unen-cumbered primary hydroxyl group in serine residues. Here, unless mild acylat-ing agents are applied, protection might indeed be indicated. Even with moder-ately active reagents an excess of the activated carboxyl derivative will cause appreciable O-acylation in the presence of a base. Such undesired O-acylation is further enhanced by imidazole and its derivatives (e.g. histidine residues in the sequence) and also by 1-hydroxybenzotriazole, often added to coupling mixtures. Thus, the serine side chain is provided with a blocking group in most syntheses. The methods used for the preparation of O-alkyl derivatives of serine and threonine are not fully satisfactory and the blocked amino acids are obtained with a certain sacrifice on time and materials. On the other hand, removal of the most commonly applied benzyl group by hydrogenation is usually quite simple, unless the catalyst is poisoned as in sulfur containing substrates. Acidolysis, however, can be the source of side reactions. Thus, deprotection with HBr in acetic acid is accompanied by partial acetylation. This can be prevented by the use of trifluoroacetic acid instead of acetic acid, but the remedy creates new problems soon to be discussed in connection with tyrosine.

The phenolic hydroxyl group in the tyrosine side chain is relatively inert and has often been left without protection even in major syntheses. Yet, the *phenolate anion* generated in the presence of base is a good nucleophile that will compete with the amino group for the acylating agent. If the latter is used in excess, then blocking of the tyrosine side chain is indeed desirable. As mentioned in the case of serine, O-acylation is promoted by the imidazole groups of histidine residues in the sequence. The frequently used benzyl group is smoothly removed by catalytic hydrogenation, but acidolysis can lead to side reaction. While HBr in acetic acid is both effective and harmless, HBr in trifluoroacetic acid yields, in addition to the parent side chain, also its 3-benzyl derivative:

Alkylation of the aromatic nucleus is due, in part, to intramolecular migration of the benzyl group, but, at least to some extent, is caused by the benzyl trifluoroacetate generated in the cleavage reaction. Interestingly, the same ring alkylation takes place even if HBr is omitted from the reagent: trifluoroacetic acid itself cleaves, albeit more slowly, the benzyl group attached to the phenolic oxygen and the C-benzyl compound is formed in a considerable part of the material. Clearly, hydrogenation and acidolysis with HBr in acetic acid are the methods of choice for the removal of the O-benzyl group from tyrosine residues but in sulfur containing peptides the former becomes inoperative and the latter can not be applied when serine occurs in the sequence. The side reaction in cleavage with HBr in trifluoroacetic acid (C-alkylation or benzyl migration) must be suppressed to make the method acceptable. Addition of phenol or cresols provides the necessary scavenging, anisole and 4-methylthiophenol are similarly effective. Benzyl migration is less pronounced with hindered blocking groups such as the cyclohexyl group. These, however require the use of very strong acids, hydrogen fluoride or trifluoromethanesulfonic acid, in the final deprotection step, reagents which can adversely affect the integrity of the assembled peptide. Thus, until a more satisfactory solution is found for the blocking of the tyrosine side chain, the choice of the protecting group must be based on the amino acid sequence of the peptide to be synthesized.

An important alternative for the benzyl group is the tert.butyl group. The tert.butyl ethers of serine, threonine and tyrosine

$$
\begin{array}{ccc}
\text{O–C(CH}_3)_3 & \text{CH}_3 & \\
| & | & \\
\text{CH}_2 & \text{CH–O–C(CH}_3)_3 & \\
| & | & \\
\text{H}_2\text{N–CH–COOH} & \text{H}_2\text{N–CH–COOH} &
\end{array}
$$

can be prepared although not without effort and sacrifice. Their use, however, is limited to syntheses in which the temporary protecting group applied for the blocking of the α-amino function is cleaved by methods other than acidolysis. In combination with the benzyloxycarbonyl group, if this can be hydrogenolyzed, or with the base-sensitive 9-fluorenylmethyloxycarbonyl (Fmoc) group for transient blocking the tert.butyl-derived masking groups are indeed of great practical value for the semipermanent protection of side chain functions.

## E. Blocking of the Sulfhydryl Group

Masking of the sulfhydryl (mercapto) function is mandatory. The nucleophilic character of the SH group is quite pronounced and can not be ignored in reactions aiming at the acylation of the amino groups. Moreover, the strong tendency of mercaptanes to be dehydrogenated to disulfides leads to mixtures, unless the operations are carried out with the exclusion of air. A simple solution for these problems is to incorporate into the peptide chain derivatives of the disulfide, cystine, instead of those of cysteine. This results, however, in doubling of the molecular-weight of the blocked intermediates and in difficulties caused by their accordingly reduced solubility in organic solvents. Therefore, this approach found only limited application in practice. A second method, blocking the SH function with the classical benzyl group, proved itself in numerous syntheses. The simple and elegant procedure for the preparation of S-benzyl-L-cysteine (du Vigneaud et al. 1930), reduction of cystine with sodium in liquid ammonia followed by alkylation *in* situ with benzyl chloride,

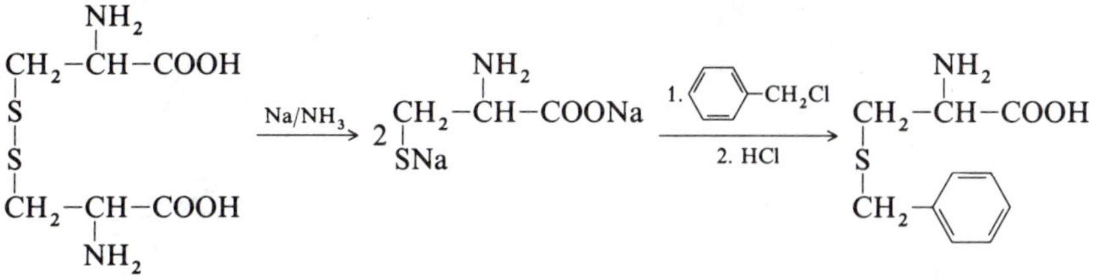

provides an inexpensive and stable intermediate for further derivatization. Also, the marked resistance of the S-benzyl group to acids allows the transient protection of α-amino groups with the benzyloxycarbonyl group because the latter can be cleaved with HBr in acetic acid without any detectable liberation of the SH group. Yet, this distinct advantage creates a new problem at the completion of chain building when the semipermanent groups are removed. Acidolytic fission of the sulfur-carbon bond is not impossible, but it requires extremely strong acidic reagents and even with liquid HF complete cleavage can be achieved only under conditions (e.g. one hour at room temperature) which endanger the integrity of certain peptide chains (cf. Chapter VII). Fortunately, reduction with sodium in liquid ammonia

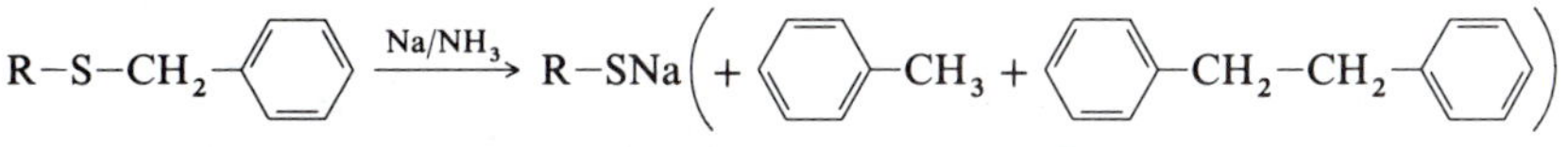

is an effective alternative that served well in many instances.

Of course, it is possible to decrease the stability of the S-benzyl group toward acids by substitution with electron-releasing groups. Thus, in the S-*p*-methoxybenzyl group the benzyl cation is sufficiently stabilized to allow cleavage of the S−C bond with liquid hydrogen fluoride under fairly mild conditions while the S-2,4,6-trimethylbenzyl group can be removed with trifluoromethanesulfonic acid in trifluoroacetic acid in the persence of *m*-cresol.

Cleavage of the S-tert.butyl group with acids is too cumbersome and is better accomplished in two steps, by treatment with *o*-nitrobenzenesulfenyl chloride followed by reduction of the thus formed mixed disulfide with mercaptoethanol:

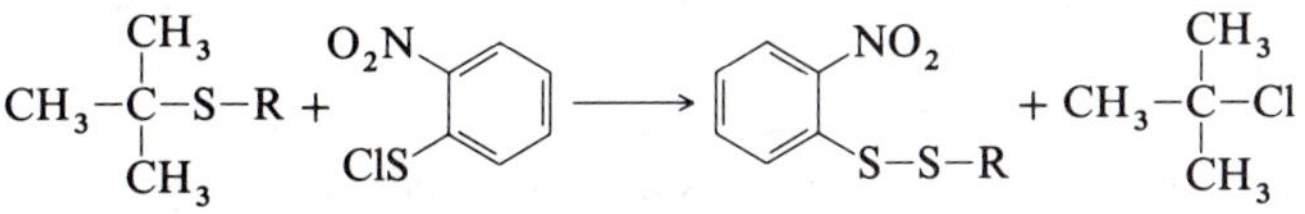

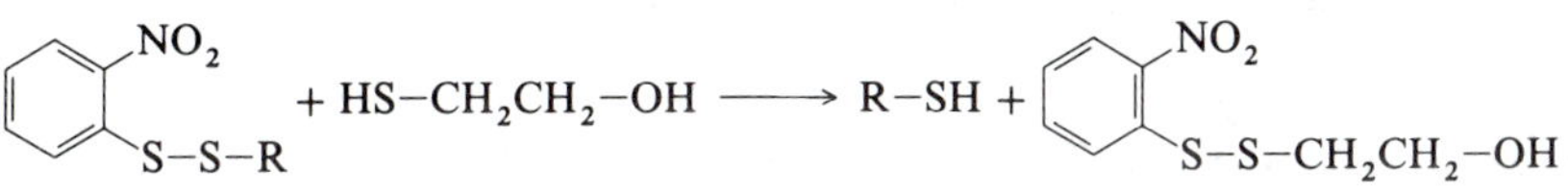

Acid sensitive blocking can be achieved by the application of the S-triphenyl-methyl (trityl) group. The S-trityl blocking is removed with HCl in organic solvents (e.g. chloroform) and also by treatment with iodine in methanol, a process which yields the disulfide:

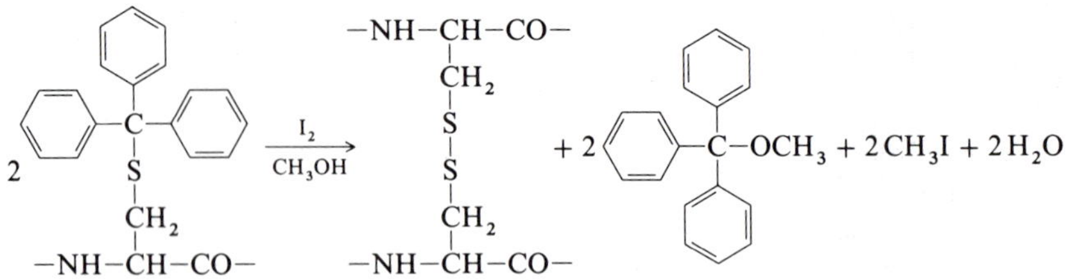

The role of methanol is quite important in this reaction; it acts as the *acceptor* of the trityl group. In a similar fashion ethylmercaptane, used in considerable excess, facilitates the acidolytic cleavage of the S-trityl group and prevents realkylation of the regenerated sulfhydryl group:

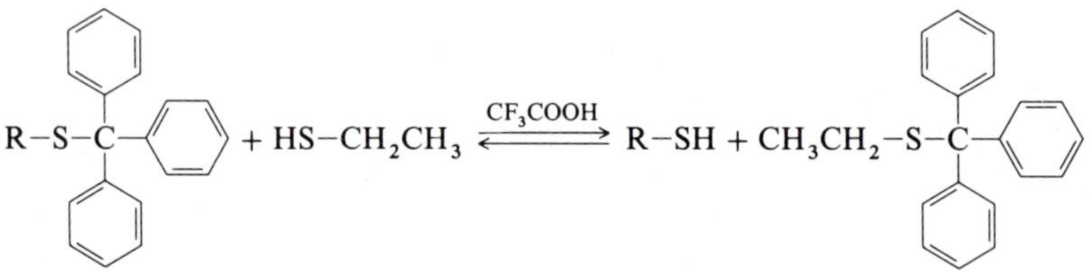

Intermediate acid sensitivities can be found in the S-diphenylmethyl (benz-hydryl) group and its methoxy derivative:

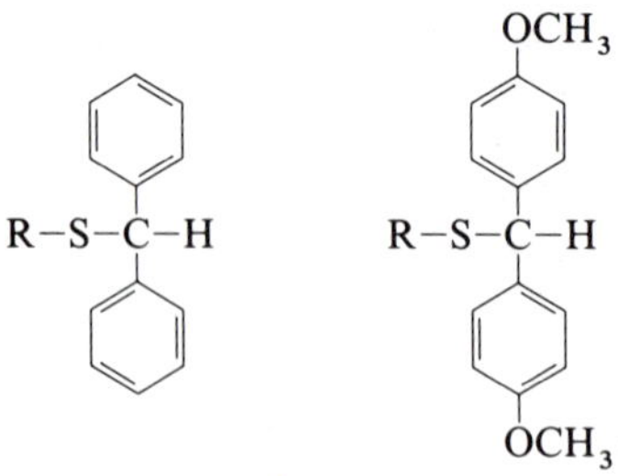

Acylation of the SH function provides no unequivocal protection, because S-acyl groups can migrate to amino groups. This S → N acyl transfer is not too pronounced in blocking with the S-benzyloxycarbonyl or with the S-tert.butyloxycarbonyl groups, but these possibilities remain to be tested in practice. Somewhat wider application has been reported of the S-ethylcarba-moyl group

$$R-S-CO-NH-CH_2CH_3$$

which, however, is fairly stable toward acids and cleaved with alkali or with bases including liquid ammonia and also hydrazine. A further possibility for its removal, treatment with mercury (II) salts, is shared with other blocking groups for the sufhydryl function, which are derived from formaldehyde. For instance the acetamidomethyl (Acm) group:

$$R-S-CH_2-NH-COCH_3$$

The Acm group gained considerable popularity, perhaps because of its alternative cleavage by iodine yielding a disulfide, as seen in the iodolysis of the S-trityl group. Deblocking the S-Acm group with $Tl(OCCF_3)_3$ follows a similar course.

From the numerous additional methods proposed for the protection of the thiol function the o-nitropyridinesulfenyl (Npys) group is mentioned here on account of the interesting reductive cleavage of the intermediate disulfide with tributylphosphine:

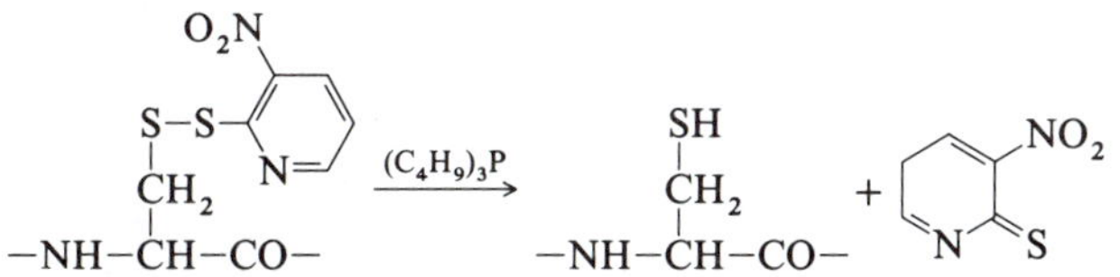

A further noteworthy although not yet exploited blocking is the alkylation of the cysteine side chain with the p-nitrobenzyl group. While it is acid resistant and can not be cleaved by hydrogenolysis, catalytic hydrogenation converts it to the p-aminobenzyl group which is then smoothly removed by salts of divalent mercury, such as $Hg^{2+}$ acetate:

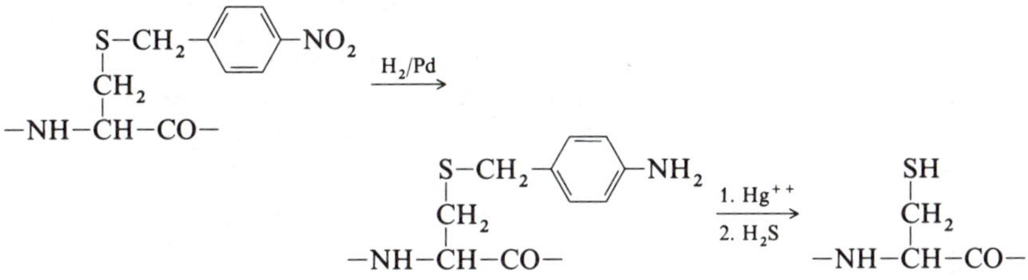

Finally the SH group has also been protected in the form of mixed disulfides, as in S-benzylmercapto-cysteine, that can be cleaved

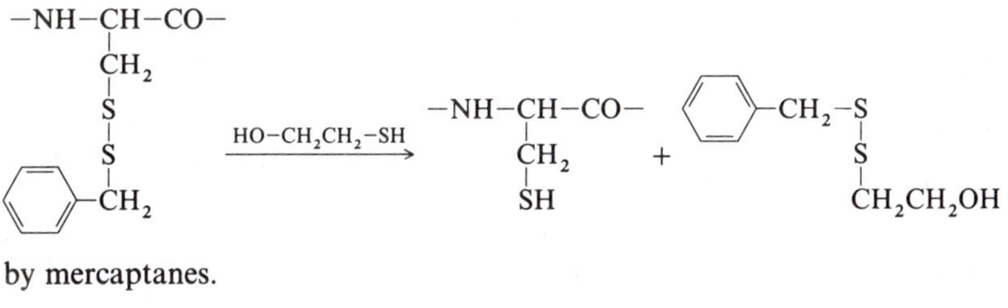

by mercaptanes.

## F. Masking the Thioether in Methionine

The thioether grouping in the side chain of methionine residues is sufficiently inert to be left unprotected during most operations of peptide synthesis. Slight oxidation to the sulfoxide can be observed in some instances, but the change is reversible: mild reducing agents, for instance thioglycolic acid, restore the sulfur atom to its original oxidation state. It is rather fortunate that in the absence of vigorous oxidizing agents no sulfone forms, because no simple reductive method can undo that damage. A more important side reaction, alkylation of the thioether sulfur, occurs in the acidolytic removal of benzyl and tert.butyl groups and in the analogous cleavage of the benzyloxycarbonyl and tert.butyloxycarbonyl groups because the thioether is an excellent acceptor of carbocations. Also, it is quite sensitive to alkylation by the by-products of acidolytic fission, namely benzyl bromide and tert.butyl trifluoroacetate:

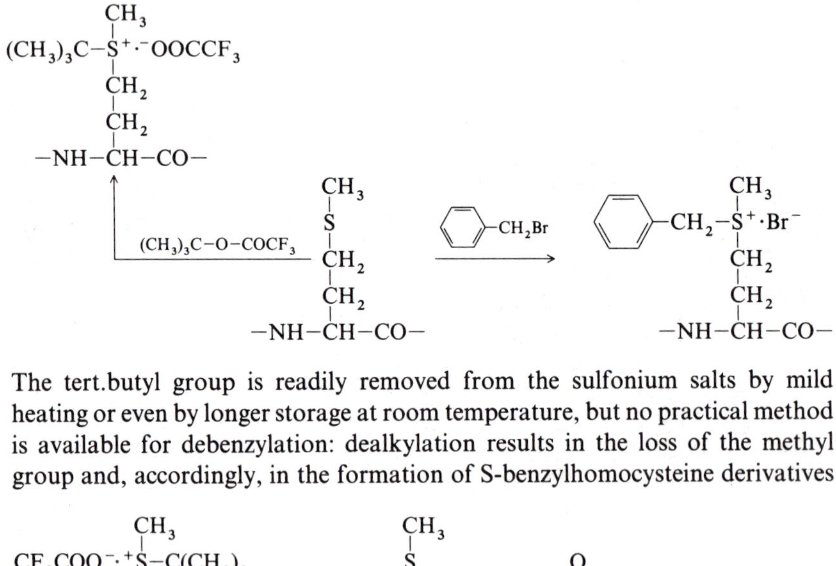

The tert.butyl group is readily removed from the sulfonium salts by mild heating or even by longer storage at room temperature, but no practical method is available for debenzylation: dealkylation results in the loss of the methyl group and, accordingly, in the formation of S-benzylhomocysteine derivatives

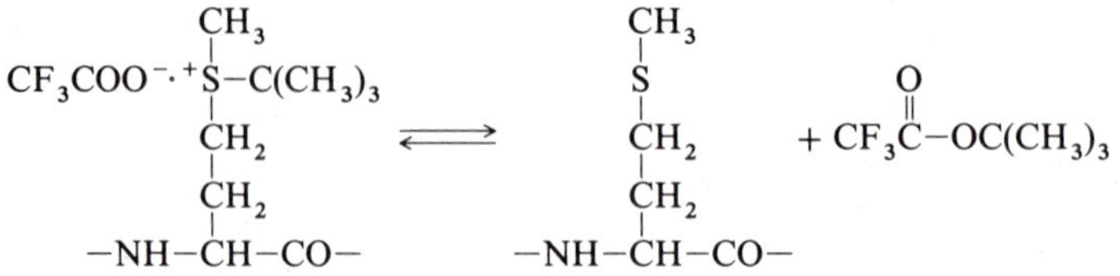

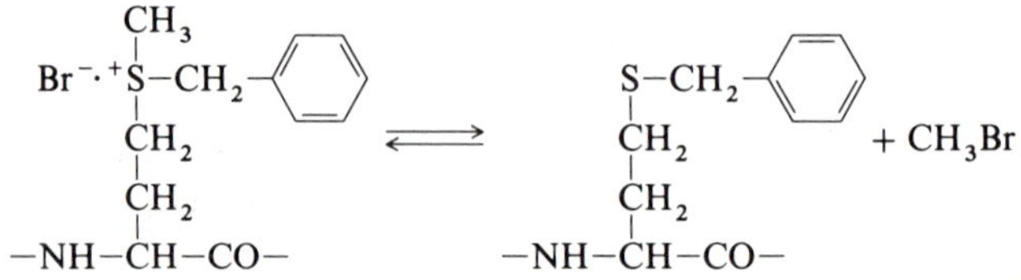

Therefore, many investigators prefer to incorporate S-blocked methionine residues into the synthetic peptide chain.

For a considerable time only one method of protection was known for this purpose, oxidation of the thioether to sulfoxide (Iselin 1961). The originally recommended oxidizing agent, peracetic acid, can be replaced by metaperiodate or by bromine water, but excellent results were obtained with aqueous hydrogen peroxide as well. At the conclusion of the chain building procedure the thioether is regenerated by reduction with thioglycolic acid, or more efficiently with N-methyl-mercaptoacetamide:

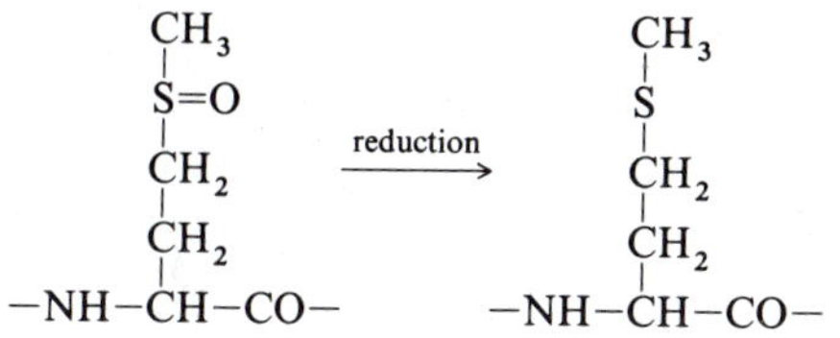

$$
\begin{array}{ccc}
\begin{array}{c}
CH_3 \\
| \\
S=O \\
| \\
CH_2 \\
| \\
CH_2 \\
| \\
-NH-CH-CO-
\end{array}
&
\xrightarrow{\text{reduction}}
&
\begin{array}{c}
CH_3 \\
| \\
S \\
| \\
CH_2 \\
| \\
CH_2 \\
| \\
-NH-CH-CO-
\end{array}
\end{array}
$$

It should be noted that the sulfur atom in methionine sulfoxide is a new chiral center, hence a pair of diastereosimers is present in the intermediates. This is a minor complication that is absent from an alternative method for the masking of the thioether: its more recently proposed conversion to a tert.sulfonium salt by the reaction of an activated N-acyl derivative of the amino acid with methyl p-toluenesulfonate:

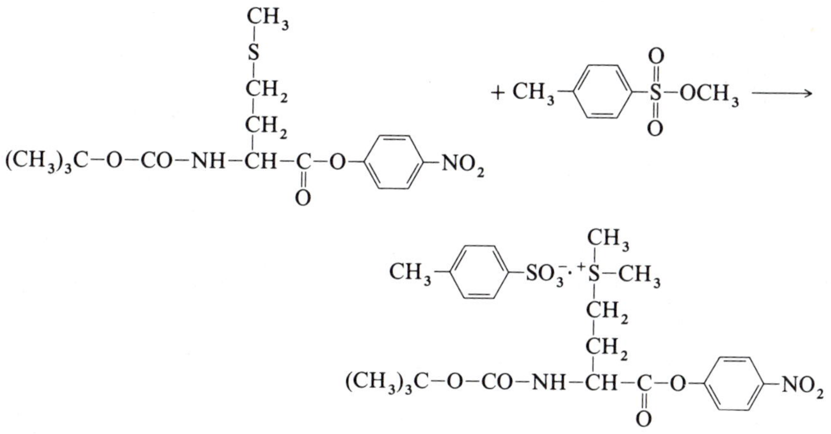

This fully protected reactive compound can be incorporated into a peptide chain and the blocking removed in the final steps by thiolysis, for instance with mercaptoethanol.

## G. Protection of the Guanidine Function in the Arginine Side Chain

Guanidine itself is an extremely strong base because its protonated form is resonance stabilized:

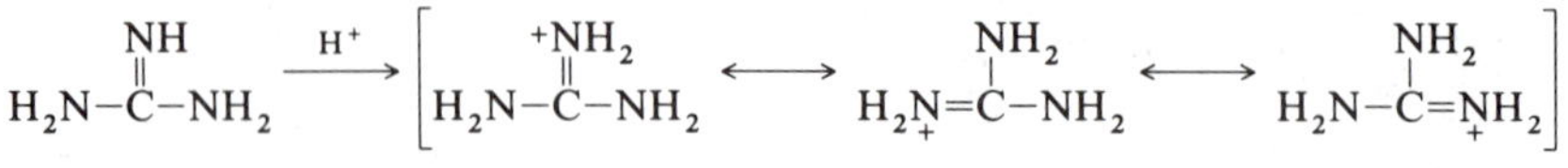

$$H_2N-\overset{\overset{NH}{\|}}{C}-NH_2 \xrightarrow{H^+} \left[ H_2N-\overset{\overset{+NH_2}{\|}}{C}-NH_2 \longleftrightarrow H_2\overset{+}{N}=\overset{\overset{NH_2}{|}}{C}-NH_2 \longleftrightarrow H_2N-\overset{\overset{NH_2}{|}}{C}=\overset{+}{N}H_2 \right]$$

The same kind of stabilization is present in monosubstitued guanidines and thus in the arginine side chain as well. The free base can be obtained with help of sodium methoxide and once formed it is subject to acylation. In protonated form, however, the guanidine group is inert toward acylating agents and can be left without further blocking. Yet, the commonly applied hydrochloride form lends unfavorable solubility properties to the intermediates. This is less pronounced in the picrate salts and, therefore, blocked and activated intermediates, such as

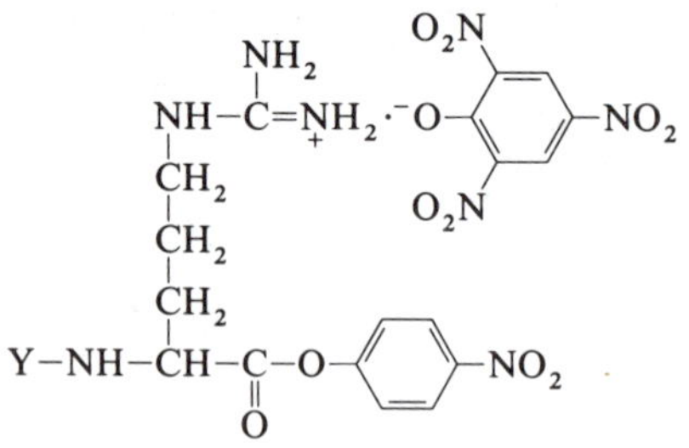

might be convenient tools for the incorporation of the arginine residue. During the various operations of chain building, however, arginine containing peptides with protonated guanidine behave like ion-echangers and in order to avoid the ensuing ambiguity many researchers prefer a more unequivocal masking of the guanidine group.

A classical approach, the use of nitroarginine, is still practiced mainly because the nitro substituent has a favorable influence on the solubility of the blocked intermediates in organic solvents. It is fairly simple to introduce the nitro group with the help of a mixture of fuming nitric acid and sulfuric acid and blocking of the α-amino group of nitroarginine is equally free of problems. It is more troublesome to activate the carboxyl group of the protected amino acid because of its pronounced tendency to produce lactams, derivatives of piperidine-2-ones:

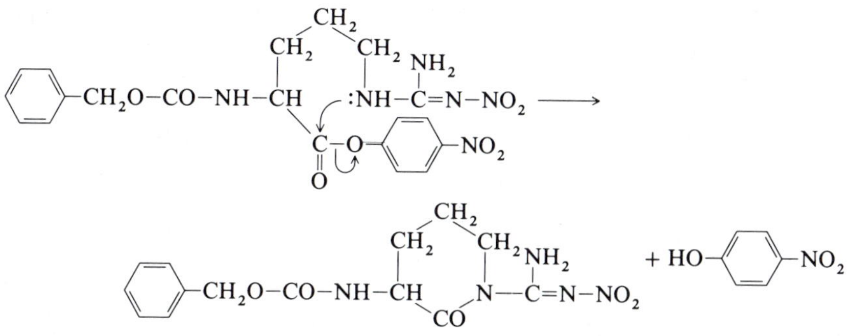

Thus, while the nitro group renders the guanidine function resistant to acylating reagents, it does not become inert toward intramolecular nucleophiles. The rate of ring closure is lower, however, in active esters with bulky activating groups

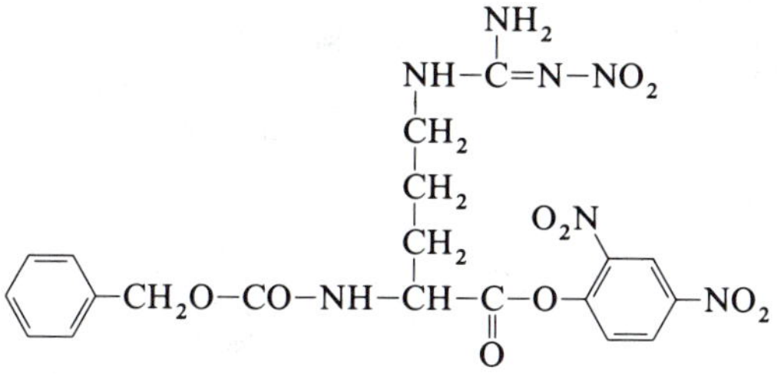

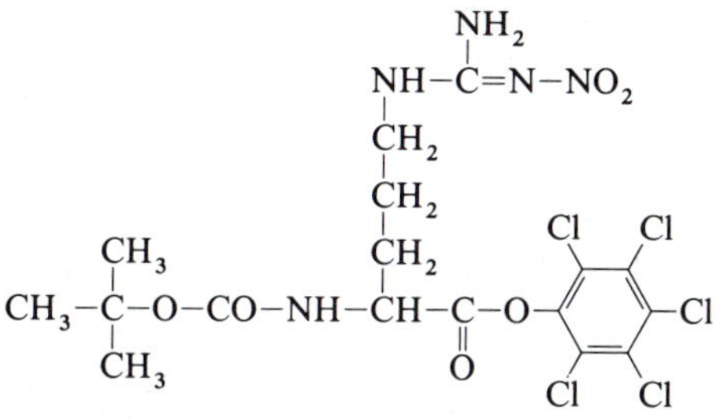

but even these are not completely stable. Therefore, activated derivatives of nitroarginine, once prepared, should be used without prolonged storage.

The initially applied method for unmasking the guanidine, catalytic hydrogenation, can be rather time consuming and extended hydrogenation might lead to saturation of aromatic side chains. Hence, a whole series of methods have been developed for the removal of the nitro group, such as catalytic reduction with cyclohexadiene as the hydrogen-donor, reduction with zinc in acetic acid, with stannous chloride or titanium(III) chloride. Acidolysis can cleave the nitro group but it resists the action of HBr in acetic acid and yields only to very strong acids, for instance liquid hydrogen fluoride or pyridinium poly-hydrogen fluoride. Additional difficulties are caused by the reduced stability of the guanidine grouping in the nitro derivative: decomposition to ornithine derivatives takes place if the intermediates are exposed to strong bases, e.g. hydrazine:

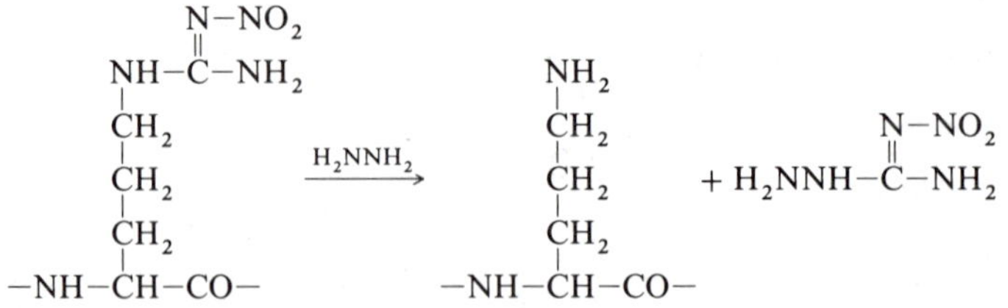

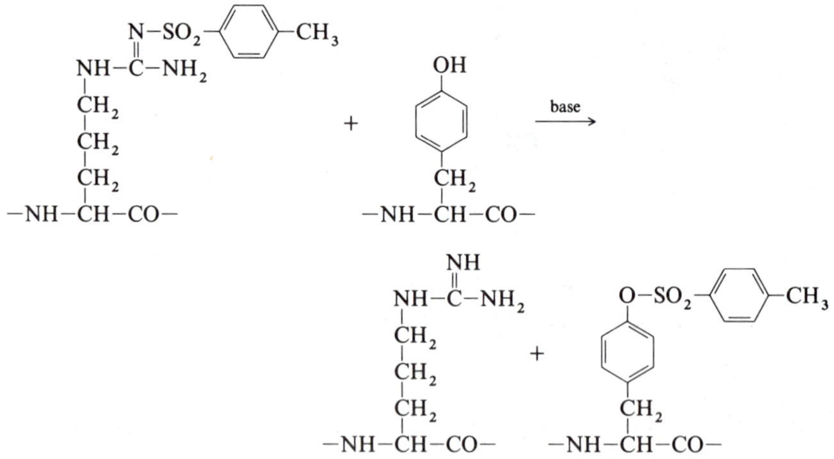

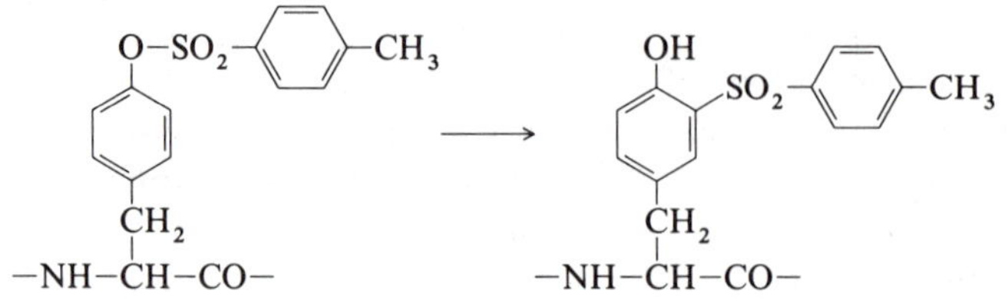

The shortcomings of the nitro group in the masking of the guanidine can not be fully eliminated by the application of the benzyloxycarbonyl or the tert.butyloxycarbonyl group for the same purpose: lactam formation and degradation to ornithine derivatives still occur. The more elaborate masking with *two* Z groups or even with *two* adamanthyloxycarbonyl groups provides only minor improvements in this respect. Acylation of the guanidine with p-toluene-sulfonyl chloride first appeared to be quite promising until the noxious transfer of the p-toluenesulfonyl (tosyl) group from the guanidine to the aromatic nucleus in tyrosine side chains was recognized. This acyl migration, that probably proceeds via O-acyl intermediates, is less pronounced if the arginine side chain is provided with the *p*-methoxybenzenesulfonyl group and still further suppressed in 2-mesitylenesulfonyl derivatives:

$$
\begin{array}{l}
\text{NH}_2 \\
| \\
\text{NH-C=N-SO}_2\text{-}\langle\bigcirc\rangle\text{-OCH}_3 \\
| \\
\text{CH}_2 \\
| \\
\text{CH}_2 \\
| \\
\text{CH}_2 \\
| \\
\text{-NH-CH-CO-}
\end{array}
\qquad
\begin{array}{l}
\text{CH}_3 \\
\text{NH}_2 \quad | \\
| \\
\text{NH-C=N-SO}_2\text{-}\langle\bigcirc\rangle\text{-CH}_3 \\
| \qquad\qquad \text{CH}_3 \\
\text{CH}_2 \\
| \\
\text{CH}_2 \\
| \\
\text{CH}_2 \\
| \\
\text{-NH-CH-CO-}
\end{array}
$$

The problems associated with the guanidine function were circumvented, in some syntheses, by the incorporation of blocked ornithine derivatives. At a late stage of the process the $\delta$-amino group of ornithine residues is deblocked and converted to the guanidino group with the help of guanylating reagents, for instance 1-guanidino-3,5-dimethylpyrazole:

$$
\begin{array}{l}
\text{NH}_2 \\
| \\
\text{CH}_2 \\
| \\
\text{CH}_2 \\
| \\
\text{CH}_2 \\
| \\
\text{-NH-CH-CO-}
\end{array}
+
\begin{array}{c}
\text{CH}_3 \\
\text{H}_2\text{N} \diagdown \\
\quad \text{C-N} \\
\text{HN} \diagup \quad \text{N} \diagdown \text{CH}_3
\end{array}
\longrightarrow
\begin{array}{l}
\text{NH}_2 \\
| \\
\text{NH-C=NH} \\
| \\
\text{CH}_2 \\
| \\
\text{CH}_2 \\
| \\
\text{CH}_2 \\
| \\
\text{-NH-CH-CO-}
\end{array}
+
\begin{array}{c}
\text{CH}_3 \\
\text{HN} \diagdown \\
\quad \text{N} \diagdown \text{CH}_3
\end{array}
$$

It seems to be fair to say that the difficulties still experienced in the synthesis of arginine containing peptides should stimulate further studies toward more satisfactory guanidine blocking groups.

# H. Masking the Imidazole in Histidine

The NH group in the imidazole nucleus of the histidine side chain has often been left unprotected with no obvious detriment to the procedure or the product of the synthesis. Some side reactions, however, can be attributed to the effect of imidazole. It acts as catalyst in the (undesired) O-acylation of serine and tyrosine side chains and also, if water is not excluded from the medium, in the hydrolysis of active esters. Furthermore, because the basic character of the imidazole is weak but not negligible, histidine containing peptides are sometimes isolated in more than one form, for instance as the base and the trifluoroacetate or as the mixture of the hydrochloride and the acetate, etc. Therefore, masking of the imidazole offers certain advantages.

Acylation of the NH group is not a real remedy because acyl-imidazoles are reactive compounds. For instance, the *p*-toluenesulfonyl group might be transferred onto other nucleophiles, such as the amino group, alcoholic or phenolic hydroxyls and quite readily to the hydroxyl group of the additive 1-hydroxybenzotriazole:

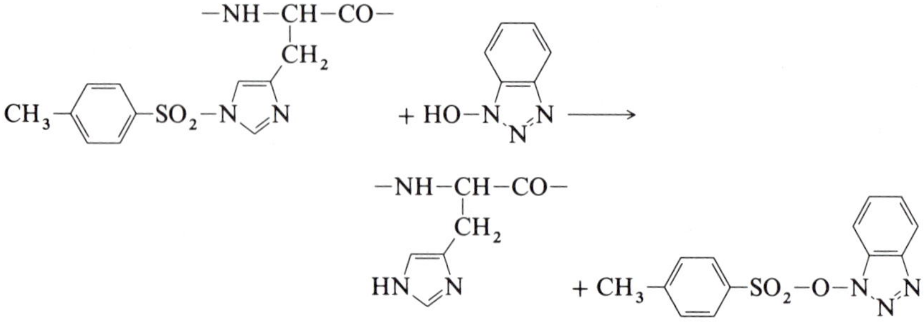

Less acyl-transfer was noted with the benzyloxycarbonyl, the tert.butyloxycarbonyl and the adamantyloxycarbonyl groups. The acid stable piperidinocarbonyl group

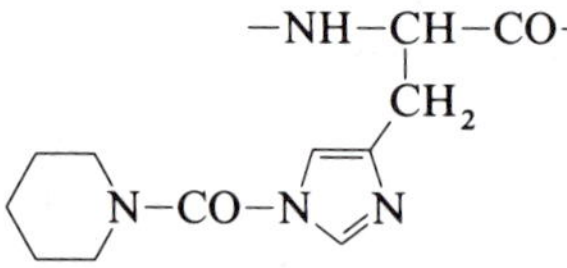

is relatively inert toward nucleophiles but can be removed with hydrazine, a supernucleophile.

Dinitrophenylation of the imidazole NH practically eliminates its basic character, yet the 2,4-dinitrophenyl (DNP) group also shares a less desirable property with real acyl groups: it can migrate to free amines:

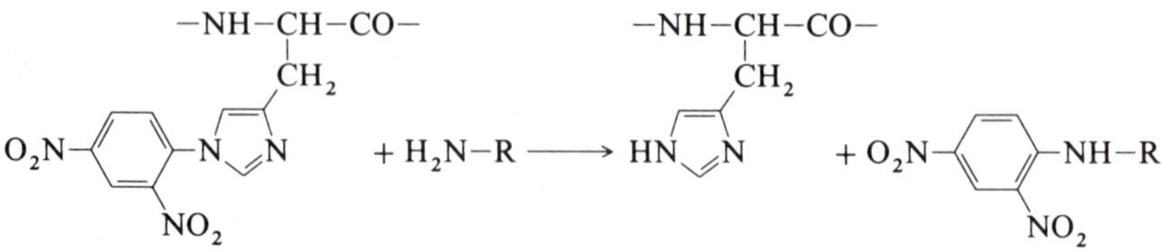

The 2,4-dinitrophenyl group is cleaved by thiols, for instance 2-mercaptoethanol, under mildly basic conditions.

The oldest method used for the protection of the histidine side chain is benzylation. The benzyl group is introduced by treatment of histidine with sodium in liquid ammonia and alkylation of the sodio derivative with benzyl chloride:

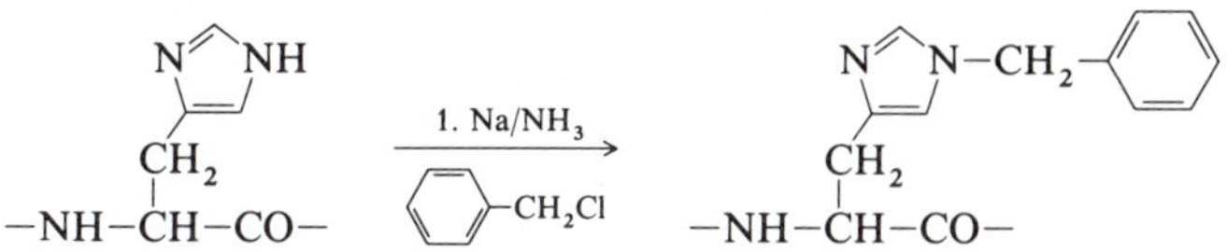

A more recently proposed approach starts with N$^{\alpha}$-benzyloxycarbonyl-histidine which is treated with benzyl bromide in the presence of a suitable organic base that plays the role of acid binding agent:

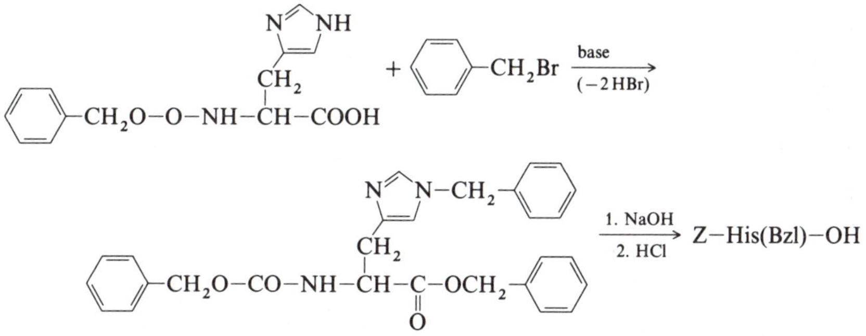

The carboxyl group is esterified in the process and has to be deblocked by saponification with alkali to allow its activation for subsequent incorporation. Deprotection of the imidazole can be achieved by catalytic hydrogenation although sometimes difficulties are encountered in this step. An alternative method of debenzylation, treatment with acetic acid, must be carried out at fairly high temperature. More acid sensitivity is shown by the im-trityl derivative while the diphenyl-4-pyridylmethyl group is acid resistant:

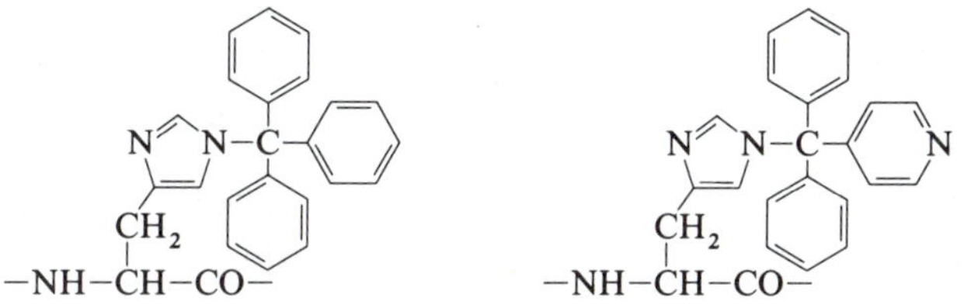

In the preceding discussion no distinction was made between the two nitrogen atoms of the imidazole ring. In the unsubstitued histidine side chain they participate in a tautomeric equilibrium and can not be readily distinguished. The situation is changed when a substituent is introduced: this can be on the $\pi$ nitrogen atom proximal to the carbon atom bonded to the CH$_2$ group of the side chain, or on the $\tau$ nitrogen which is further removed from the point of attachment. Thus two im-benzyl derivatives are possible, $\pi$-benzyl and $\tau$-benzyl histidine

In the structures drawn so far it was tacitly assumed that the less hindered nitrogen carries the blocking group and this assumption is probably justified in the case of bulky substituents. Because of the role of the $\pi$-nitrogen in the sometimes not negligible racemization of histidine during coupling, a few attempts were made for the regioselective preparation of distinctly N$^\pi$-blocked derivatives. The significance of these compounds, for instance N$^\alpha$-benzyloxy-carbonyl-N$^\pi$-phenacyl-L-histidine and N$^\alpha$-tert.butyloxycarbonyl-N$^\pi$-benzyl-oxymethyl-L-histidine

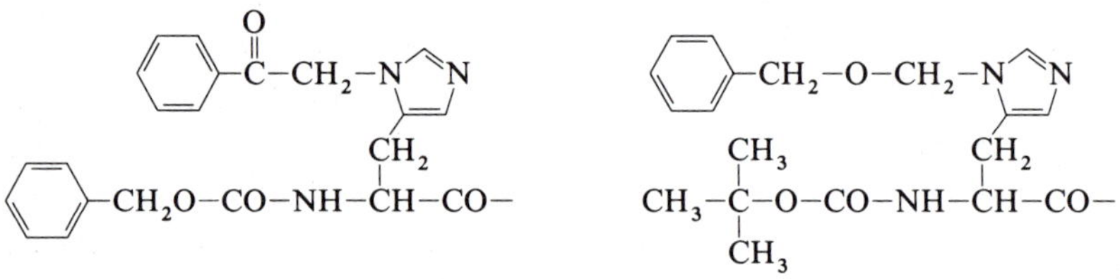

still has to be established through their application in major syntheses.


# I. Indole Protection


The indole nucleus in the tryptophan side chain is unreactive enough to be left without protection. Its NH group is not readily acylated but can suffer alkylation to a small but not always negligible extent and some investigators prefer to counteract this side reaction by the application of a masking group. So far only the formyl group found acceptance, perhaps because it is smoothly introduced by the treatment of tryptophan with HCl in formic acid:

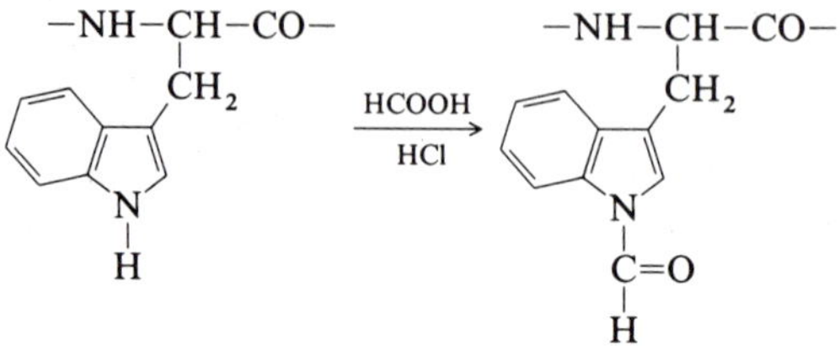

The in-formyl group is cleaved by weak bases such as aqueous piperidine or hydrazine. Its sensitivity to amines can pose some problem: in mildly basic media it migrates to $\alpha$-amino groups. On the other hand, the in-formyl group is indeed able to reduce the extent of alkylation, for instance tert.butylation that occurs during the removal of tert.butyl groups by acidolysis. Also, in $N^{in}$-formyl derivatives of tryptophan containing peptides less oxidation takes place in acidic media, a reaction that results in the formation of colored products.

In recent years attempts were made to mask the indole nitrogen with the benzyloxycarbonyl group and this became possible through the use of "naked" fluoride anions generated with crown ethers. The benzyloxycarbonyl group can be removed from the indole nucleus by the action of strong acids, for instance liquid hydrogen fluoride and also by catalytic hydrogenation. In spite of its inherent promise no general application of the in-Z group can be found in the literature so far.


## J.  Carboxamide Blocking Groups


The side chain carboxamide groups of asparagine and glutamine can lose water during activation. There are several ways to avoid this side reaction. For instance, no significant amount of nitrile forms if the lifetime of the reactive intermediate is cut short by the addition of an auxiliary nucleophile such as 1-hydroxybenzotriazole (HOBt):

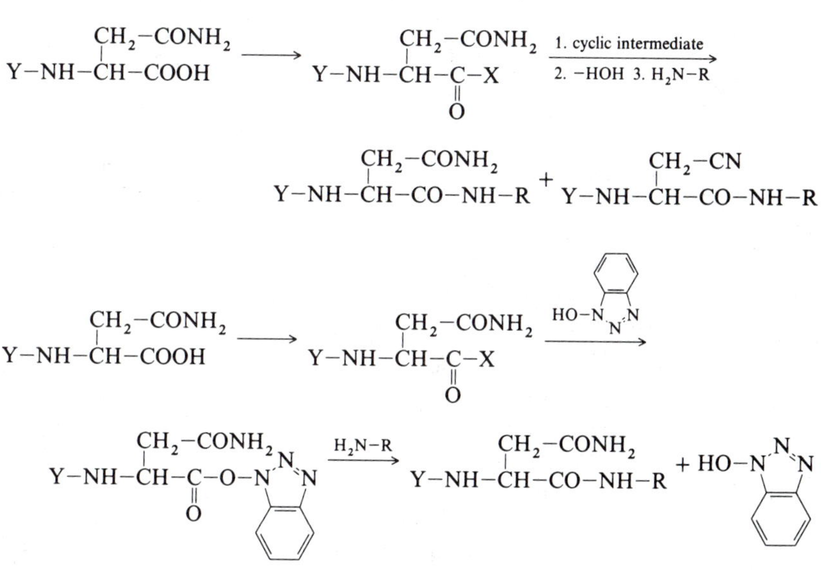

An alternative solution for the problem is the preparation of active esters of ($N^{\alpha}$-blocked) asparagine and glutamine and to separate the desired carboxamide derivative from the nitrile. The purified reactive intermediates

$$
\begin{array}{c}
CH_2CONH_2 \\
| \\
Z-NH-CH-COOH
\end{array}
+ HO-\!\!\left\langle\bigcirc\right\rangle\!\!-NO_2 \xrightarrow{\;\bigcirc\!-N=C=N-\bigcirc\;}
$$

$$
\begin{array}{c}
CH_2CONH_2 \\
| \\
Z-NH-CH-C-O-\!\!\left\langle\bigcirc\right\rangle\!\!-NO_2 \\
\quad\quad\quad\; \| \\
\quad\quad\quad\; O
\end{array}
+
\begin{array}{c}
CH_2CN \\
| \\
Z-NH-CH-C-O-\!\!\left\langle\bigcirc\right\rangle\!\!-NO_2 \\
\quad\quad\quad\; \| \\
\quad\quad\quad\; O
\end{array}
$$

are sufficiently stable to be stored and can be used for the incorporation of asparagine or glutamine residues. Because of these remedial possibilities blocking of the carboxamide group is relatively unimportant. If, however, it appears necessary, a whole series of masking groups is available. The classical method, condensation of the amide with xanthydrol, is not ideal, because the substituted intermediates are less soluble in organic solvents than the unprotected ones and also, because removal of the xanthydryl group requires fairly vigorous acidolysis. The benzhydryl group is similar in this respect while its dimethoxy-derivative is more sensitive to acids.

$$
\begin{array}{ccc}
\begin{array}{c}
\text{(xanthydryl)} \\
CO-NH\ H \\
| \\
CH_2 \\
| \\
-NH-CH-CO-
\end{array}
&
\begin{array}{c}
CO-NH-\overset{\bigcirc}{\underset{\;}{C}}-H \\
| \\
CH_2 \\
| \\
CH_2\ \bigcirc \\
| \\
-NH-CH-CO-
\end{array}
&
\begin{array}{c}
OCH_3 \\
CO-NH-\overset{\bigcirc OCH_3}{\underset{\;}{C}}-H \\
| \\
CH_2 \\
| \\
CH_2 \\
| \\
-NH-CH-CO-\ OCH_3
\end{array}
\end{array}
$$

The 2,4-dimethoxy- and 2,4,6-trimethoxy-benzyl groups

$$
\begin{array}{cc}
\begin{array}{c}
CH_3O \\
CO-NH-CH_2-\!\!\left\langle\bigcirc\right\rangle\!\!-OCH_3 \\
| \\
CH_2 \\
| \\
CH_2 \\
| \\
-NH-CH-CO-
\end{array}
&
\begin{array}{c}
CH_3O \\
CO-NH-CH_2-\!\!\left\langle\bigcirc\right\rangle\!\!-OCH_3 \\
| \quad\quad\quad\quad\quad CH_3O \\
CH_2 \\
| \\
CH_2 \\
| \\
-NH-CH-CO-
\end{array}
\end{array}
$$

are cleaved by trifluoroacetic acid but not under mild conditions. These difficulties are the reason for the incorporation of asparagine and glutamine residues without side chain protection in most instances.

# References

Barton, M. A., Lemieux, R. U., Savoie, J. Y.: J. Amer. Chem. Soc. *95,* 4501 (1973)
Ben Ishai, D., Berger, A.: J. Org. Chem. *17,* 1564 (1952)
Bergmann, M., Zervas, L.: Ber. dtsch. Chem. Ges. *65,* 1192 (1932)
Carpino, L. A., Han, G. Y.: J. Amer. Chem. Soc. *92,* 5748 (1970)
Carpino, L. A., Tsao, J. H., Ringsdorf, H., Fell, E., Hettich, G.: J. Chem. Soc. Chem. Commun. *1978,* 358
du Vigneaud, V., Audrieth, L. F., Loring, H. S.: J. Amer. Chem. Soc. *52,* 4500 (1930)
Fischer, E.: Ber. dtsch. Chem. Ges. *24,* 239 (1902)
Iselin, B.: Helv. Chim. Acta *44,* 61 (1961)
Sieber, P., Iselin, B.: Helv. Chim. Acta *51,* 614, 622 (1968)
Yajima, H., Takeyama, M., Kanaki, J., Mitani, K.: J. Chem. Soc. Chem. Commun. *1978,* 482

# Additional Sources

Geiger, R., König, W.: Amine Protecting Groups, in The Peptides, vol. 3, Gross, E., Meienhofer, J., eds., pp. 1–99, New York, Academic Press 1979
Hiskey, R. G.: Sulfhydryl Protecting Groups, in The Peptides, vol. 3, Gross, E., Meienhofer, J., eds., pp. 137–167, New York, Academic Press 1979
Roeske, R. W.: Carboxyl Protecting Groups, in The Peptides, vol. 3, Gross, E., Meienhofer, J., eds., pp. 101–136, New York, Academic Press 1979

# VII. Undesired Reactions During Synthesis

It might be nigh impossible to reduce peptide synthesis to a mere routine, because the intended transformations, such as blocking, activation, coupling and removal of protecting groups, are accompanied by numerous undesired reactions. Therefore, instead of a single product often a mixture of peptides is obtained and the target compound must be secured through purification, that may require several steps and sometimes tedious, time consuming operations. Consequently, the final yield on homogeneous material can be rather disappointing. Thus, it is absolutely necessary to recognize and to understand such undesired reactions, to anticipate and, when possible, to prevent them. Here we can discuss only those side reactions which are repeatedly encountered and reported. An important problem, racemization during synthesis will be treated in a separate chapter.

## A. Base Catalyzed Ring Closure

In the process of chain-lengthening addition of the third residue (or a peptide) to a *dipeptide derivative* is an ambiguous step. Alkyl esters of dipeptides undergo spontaneous cyclization and produce 2,5-diketopiperazines:

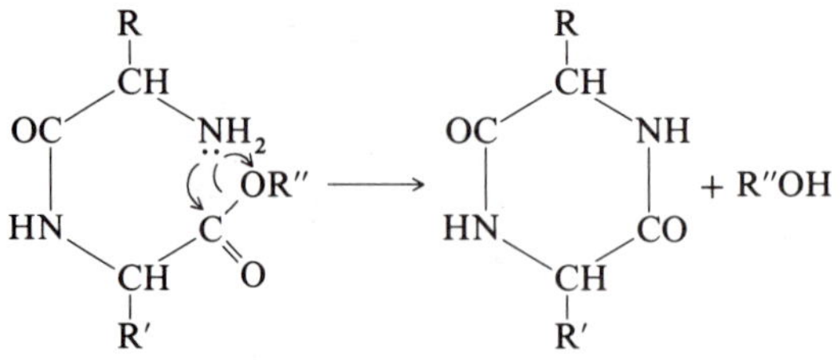

In diketopiperazines (DKP-s) both amides are *cis,* that is the carbonyl oxygen atom and the amide hydrogen are on the same side of the $C-N$ partial double bond. Accordingly, cyclization requires the energy consuming conversion of the more stable transoid to the less stable cisoid form. The energy liberated in the formation of the very stable diketopiperazine structure provides the necessary driving force for the reaction:

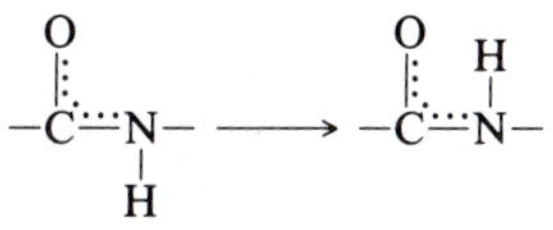

Ring formation is particularly pronounced in glycine and in proline containing peptides: in the former because of the absence of an interfering side chain, in the latter because the cyclic side chain lies in the plane of the DKP molecule and is out of the way during ring closure. Similarly, diketopiperazines are more readily formed when one of the residues belong to the L, the other to the D family of amino acids. In this case the side chains are on opposite sides of the general plane of the DKP while if both amino acids have the same configuration then both side chains are on the same side and their bulk interferes with ring closure.

Diketopiperazine formation readily occurs in methyl, ethyl and benzyl esters of dipeptides, but not in tert.butyl esters. In dipeptide amides it takes place only at elevated temperatures. High concentration of the reactants enhance the rate of the desired coupling reaction and thus reduce the extent of cyclization, which is intramolecular and hence independent of concentration. Storage of dipeptide esters having a free amino group should, of course, be avoided. Yet, diketopiperazine derivatives might be generated also from dipeptides in which the amino group is blocked. For instance, benzyloxycarbonyl-glycyl-proline p-nitrophenyl ester yields, in the presence of a base an acyl-diketopiperazine:

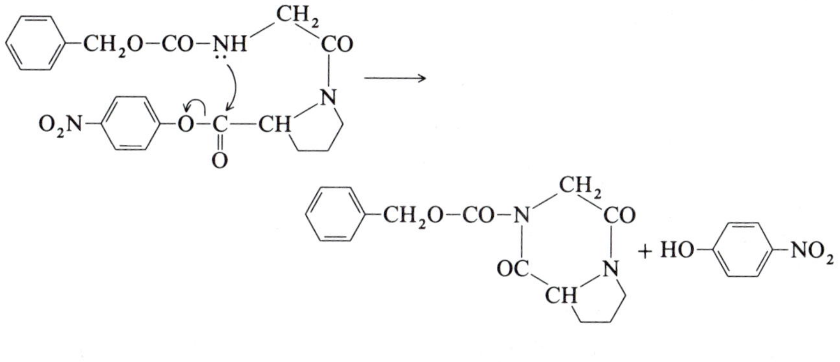

A different base catalyzed cyclization was observed during the saponification of peptide esters in which the N-terminal residue was blocked by the benzyloxycarbonyl group:

$$\langle\bigcirc\rangle\!-\!CH_2O\!-\!CO\!-\!NH\!-\!CHR\!-\!CO\!-\!NH\!-\!CHR'\!-\!CO\!-\!OR'' \xrightarrow{OH^-}$$

CHR—CO                           CHR—CO
HN          N—CHR'—COO⁻ ⟶ HN          N—CHR'—COO⁻ + ⟨◯⟩—CH₂O⁻
    C=O                              CO
    C–O
    CH₂
    ⟨◯⟩

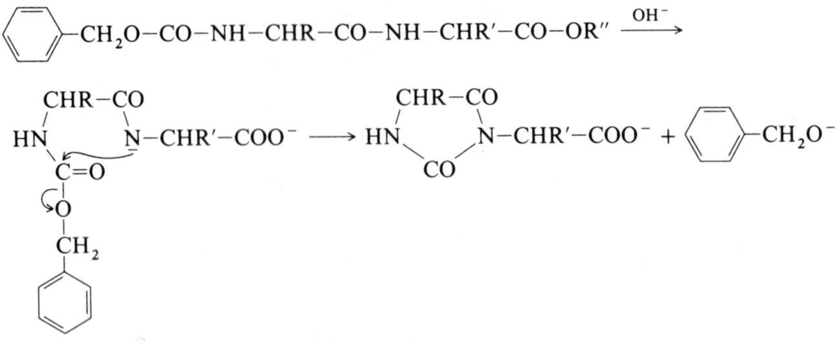

The cyclization products (hydantoins) are cleaved by excess alkali to give urea derivatives, dicarboxylic acids with the structure

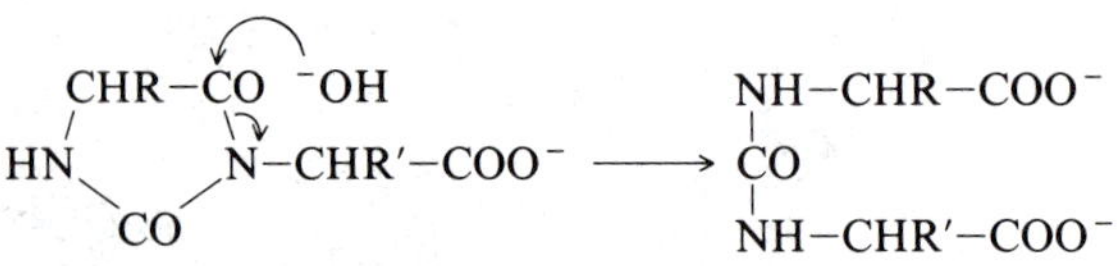

CHR—CO ⁻OH              NH—CHR—COO⁻
HN      N—CHR'—COO⁻ ⟶ CO
    CO                     NH—CHR'—COO⁻

This side reaction can become quite significant if R' is hydrogen, that is when glycine is the second residue in the sequence, because glycine is the only amino acid that can readily by di-acylated at the amino group.

An analogous ring closure to a hydantoin derivative occurs during the ammonolysis of benzyloxycarbonyl-L-phenylalanyl-glycine ethyl ester:

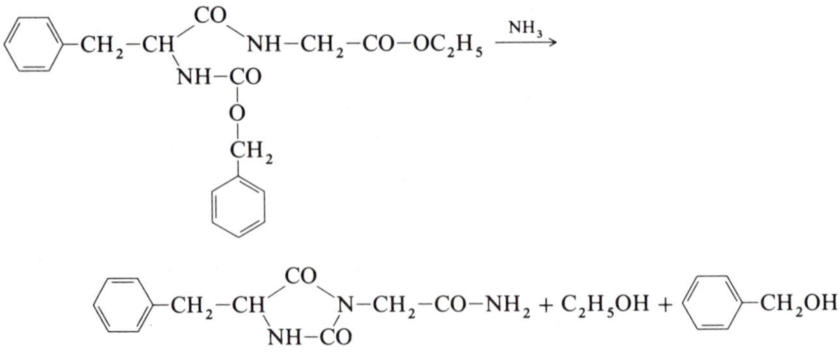

$$\langle\bigcirc\rangle\!-\!CH_2\!-\!CH \overset{CO}{\underset{NH-CO}{\diagdown}} NH\!-\!CH_2\!-\!CO\!-\!OC_2H_5 \xrightarrow{NH_3}$$

                              O
                              CH₂
                              ⟨◯⟩

$$\langle\bigcirc\rangle\!-\!CH_2\!-\!CH \overset{CO}{\underset{NH-CO}{\diagdown}} N\!-\!CH_2\!-\!CO\!-\!NH_2 + C_2H_5OH + \langle\bigcirc\rangle\!-\!CH_2OH$$

Base catalyzed formation of hydantoins is not limited to blocked dipeptides, it can take place in derivatives of longer chains as well if they are blocked with the benzyloxycarbonyl group at the N-terminus and contain glycine as the second residue in their sequence.

Proton abstraction from an amide nitrogen initiates a similar inramolecular nucleophilic attack in peptides which contain aspartyl residues with the side chain carboxyl group blocked in the form of benzyl ester:

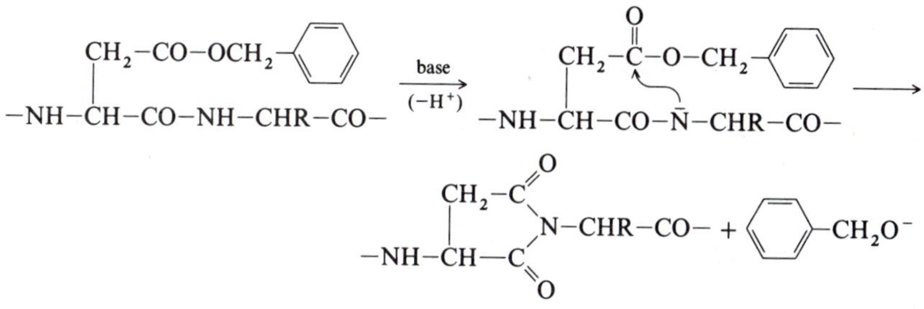

$$
\underset{\substack{|\\ -NH-CH-CO-NH-CHR-CO-}}{CH_2-CO-OCH_2-\bigcirc} \quad \xrightarrow[(-H^+)]{base} \quad \underset{\substack{|\\ -NH-CH-CO-\overset{..}{N}-CHR-CO-}}{CH_2-\overset{\overset{\displaystyle O}{\|}}{C}-O-CH_2-\bigcirc} \quad \longrightarrow
$$

$$
\underset{\substack{|\\ -NH-CH-C}}{CH_2-C}\underset{\diagdown\\O}{\overset{\diagup O}{\diagdown}}N-CHR-CO- \quad + \quad \bigcirc-CH_2O^-
$$

The rate of ring closure is a function of the sequence; it is fast if R = H, that is in peptides containing the — Asp-Gly — partial sequence and orders of magnitude slower if the aspartyl residue is followed by an amino acid with a bulky side chain, such as valine (R = isopropyl). Opening of the succinimide ring with alkali is not helpful because β-aspartyl peptides are produced as the preponderant product:

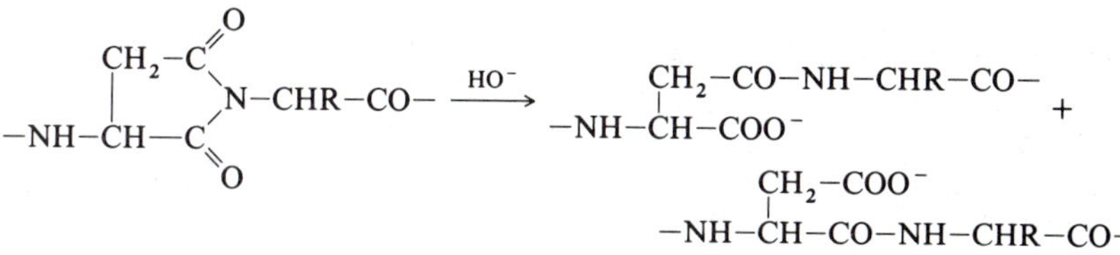

$$
\underset{\substack{|\\-NH-CH-C}}{CH_2-C}\;N-CHR-CO- \quad \xrightarrow{HO^-} \quad \underset{\substack{|\\-NH-CH-COO^-}}{CH_2-CO-NH-CHR-CO-} \quad +
$$

$$
\underset{\substack{|\\-NH-CH-CO-NH-CHR-CO-}}{CH_2-COO^-}
$$

While with β-benzyl aspartyl residue succinimide formation can be quite extensive, the corresponding tert.butyl ester are fairly inert toward intramolecular attack.

The corresponding cyclization of glutamyl residues is mostly insignificant, but glutarimides (piperidine-2,5-diones) are generated from carboxyl-activated N-acyl-glutamines:

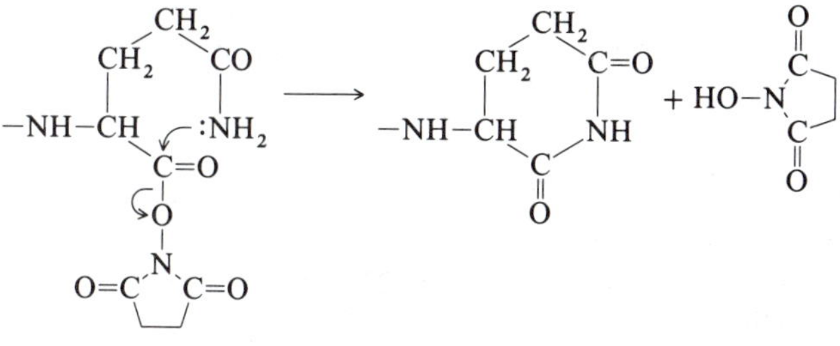

$$
\underset{\substack{|\\ -NH-CH\\ \diagdown C=O\\ \diagdown O\\ |\\ N\\ O=C\diagup\diagdown C=O}}{\overset{CH_2}{\underset{CH_2}{\diagup}\diagdown CO}} \quad \longrightarrow \quad \underset{\substack{|\\ -NH-CH\quad NH\\ \diagdown C\diagup\\ \|\\ O}}{\overset{CH_2}{\underset{CH_2}{\diagup}\diagdown C=O}} \quad + HO-N\underset{\substack{C\\ \|\\ O}}{\overset{\overset{\displaystyle O}{\|}}{C}}
$$

The alternative ring closure yielding derivatives of pyroglutamic acid (a pyrrolidone) also can occur, for instance, when p-toluenesulfonyl-glutamic acid dichloride is stored or exposed to the action of base:

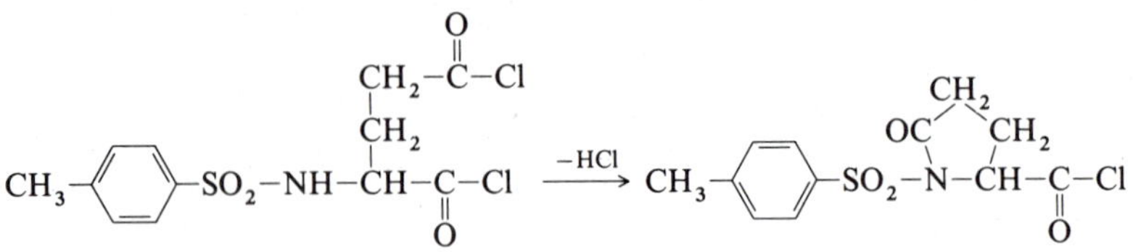

## B. Acid Catalyzed Ring Closure

Under the influence of strong acids β-benzyl aspartyl residues give rise to aminosuccinyl peptides

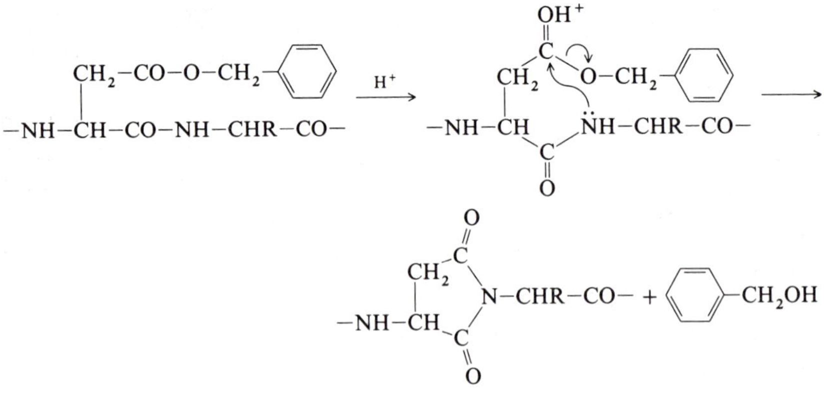

and similar cyclization can occur even when the β-carboxyl group is free. In fact the same side reaction has been encountered under essentially neutral conditions as well.

In peptides with glutamine as the N-terminal residue spontaneous pyrrolidone formation takes place; the reaction is catalyzed by weak acids:

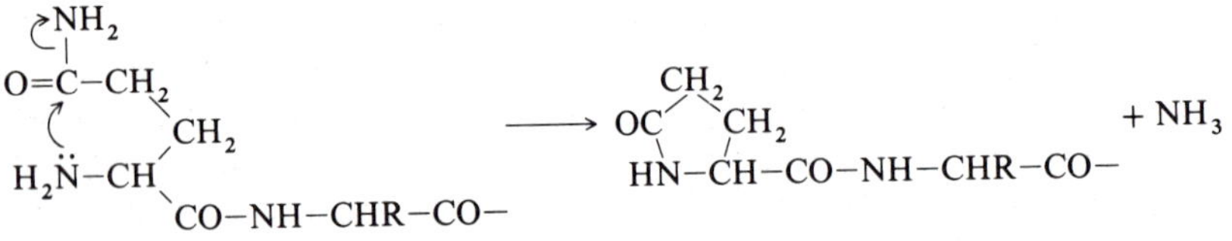

# C. Acyl Migration

In addition to base catalyzed O-acylation of hydroxyl groups in serine, threonine and tyrosine side chains (cf. Chapter VI), an acid catalyzed side reaction involving serine residues needs to be mentioned. Fortunately it does not occur under the relatively mild acidic conditions that prevail during the removal of acid-labile $\alpha$-amine protecting groups, but in some very strong acids used for final deprotection the peptide chain portion attached to the amino group of a serine residue migrates to its hydroxyl group:

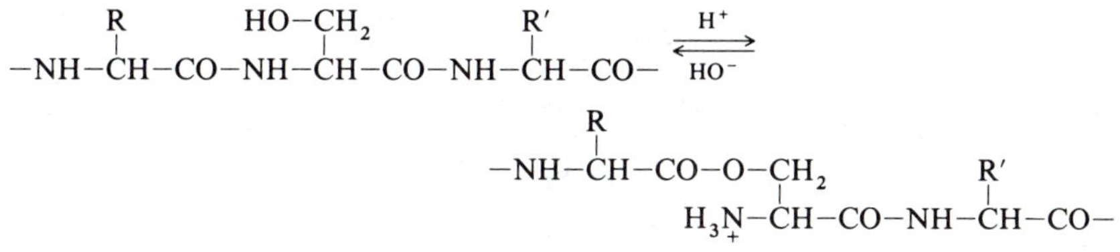

$$
\begin{array}{ccccc}
R & HO-CH_2 & R' & H^+ \\
-NH-CH-CO-NH-CH-CO-NH-CH-CO- & \underset{HO^-}{\rightleftharpoons}
\end{array}
$$

$$
\begin{array}{cc}
R \\
-NH-CH-CO-O-CH_2 & R' \\
H_3\overset{+}{N}-CH-CO-NH-CH-CO-
\end{array}
$$

As indicated above, the reaction is reversed in basic solution, for instance in aqueous bicarbonate. This reversal, however, is accompanied by partial hydrolysis of the newly formed ester bond and hence the damage can not be entirely undone. Practically no N → O migration takes place in liquid HF at 0 °C in about 20 minutes. Yet, when nitro groups must be removed from guanidine moieties or S-benzyl groups need to be cleaved then exposure to HF at room temperature and a reaction time as long as one hour is required. In such instances N → O migration becomes a disturbing reality. Therefore, in the synthesis of serine containing peptides it would seem advisable to select protection schemes that do not necessitate drastic acidolysis in the concluding steps.

# D. Alkylation

Electrophilic substitution of aromatic nuclei in tyrosine and tryptophan side chains has frequently been reported in connection with acidolytic removal of blocking groups. C-Benzylation and tert.butylation of the tyrosine side chain and N-alkylation of the indole nucleus in tryptophan are often attributed to the alkyl cations generated in the reaction. This common side reaction is caused, however, mainly by the alkylating agents formed in the process, such as benzyl bromide or tert.butyl trifluoroacetate. The same is true for the S-alkylation of the methionine side chain. Conversion of the thioether to a sulfonium salt can occur for an additional reason as well. In strong acids anisole, added as cation-scavanger, transfers its methyl group to the sulfur atom:

$$\begin{array}{c} CH_3 \\ | \\ S \\ | \\ CH_2 \\ | \\ CH_2 \\ | \\ -NH-CH-CO- \end{array} \quad + \quad \underset{}{\bigcirc}-OCH_3 \quad \underset{}{\overset{HF}{\rightleftharpoons}} \quad \begin{array}{c} CH_3 \\ | \\ CH_3-\overset{+}{S}\cdot F^- \\ | \\ CH_2 \\ | \\ CH_2 \\ | \\ -NH-CH-CO- \end{array} \quad + \quad \underset{}{\bigcirc}-OH$$

This side reaction is reversible. As already mentioned in connection with the protection of the methionine side chain (Chapter VI) the thioether can be regenerated from the ternary sulfonium salts by treatment with mercaptanes under mildly basic conditions. Nevertheless, this long overlooked alkylation should serve as warning against the indiscriminate use of scavangers. The more acid stable thioanisole or the equally efficient 4-methylthiophenol are less likely to cause similar problems.

# E. Friedel-Crafts Reaction

Anisole, implicated in the preceding paragraph, can be the cause of an additional side reaction, another electrophilic substitution catalyzed by strong acids. To wit, the side chain carboxyl group of glutamyl residues participates in the Friedel-Crafts acylation of the scavenger and yields a stable ketone:

$$\begin{array}{c} CH_2-COOH \\ | \\ CH_2 \\ | \\ -NH-CH-CO- \end{array} \quad + \quad \underset{}{\bigcirc}-OCH_3 \quad \overset{HF}{\longrightarrow} \quad \begin{array}{c} CH_2-\overset{O}{\overset{||}{C}}-\underset{}{\bigcirc}-OCH_3 \\ | \\ CH_2 \\ | \\ -NH-CH-CO- \end{array}$$

It is similarly difficult to prevent the Friedel-Crafts reaction between side chain carboxyl groups and the aromatic nuclei of insoluble polymeric supports used in solid phase peptide synthesis (cf. Chapter X). Here, however, the ketone by-products are firmly anchored to the polymer:

$$\begin{array}{c} CH_2-COOH \\ | \\ CH_2 \\ | \\ -NH-CH-CO- \end{array} \quad + \quad \begin{array}{c} | \\ CH_2 \\ | \\ \underset{}{\bigcirc}-CH \\ | \\ CH_2 \\ | \end{array} \quad \overset{HF}{\longrightarrow} \quad \begin{array}{c} CH_2-\overset{O}{\overset{||}{C}}-\underset{}{\bigcirc}-\begin{array}{c} | \\ CH_2 \\ | \\ CH \\ | \\ CH_2 \\ | \end{array} \\ | \\ CH_2 \\ | \\ -NH-CH-CO- \end{array}$$

Therefore, while they represent a loss, no ketonic material contaminates the peptide cleaved from the resin.

## F. Saturation of Aromatic Nuclei

Removal of the benzyloxycarbonyl group by catalytic hydrogenation and hy-
drogenolytic cleavage of benzyl esters or benzyl ethers is usually a fast process
and the conditions applied (palladium catalyst, atmospheric pressure) are not
conducive to significant saturation of aromatic rings. A more energetic procedure, for instance the use of platinum in acetic acid, can be more risky. Yet, even
under the otherwise harmless conditions of hydrogenolysis, partial saturation
of phenylalanine and tryptophan residues have been noted when the reaction
had to be extended to several days. This can become necessary when intermediates containing nitroarginine and/or im-benzyl-histidine residues are deprotected by catalytic hydrogenation. Prolonged catalytic reduction was intention-
ally applied for the preparation of hexahydrophenylalanine (cyclohexylalanine)
containing peptides. It must be noted, however, that hydrogenation of tyrosine
gives rise both to hexahydrotyrosine and hexahydrophenylalanine.

## G. Reductive Cleavage of the Peptide Chain

When blocking groups are cleaved by reduction with sodium in liquid ammonia
the peptide bonds are generally not affected. An exception is found in proline
containing peptides. The *secondary* amide in the acyl-proline residue undergoes
reductive fission to yield an aldehyde:

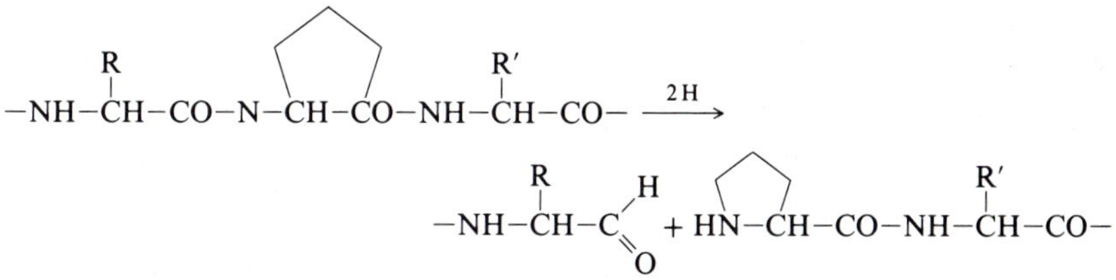

$$-NH-\overset{R}{\underset{|}{CH}}-CO-N-\overset{\triangle}{\underset{|}{CH}}-CO-NH-\overset{R'}{\underset{|}{CH}}-CO- \xrightarrow{2H}$$

$$-NH-\overset{R}{\underset{|}{CH}}-C\overset{H}{\underset{O}{}} + HN-CH-CO-NH-\overset{R'}{\underset{|}{CH}}-CO-$$

This reaction requires the presence of a hydrogen donor, such as water or
methanol and does not occur if anhydrous conditions are maintained.

## H. Oxidative Decomposition of the Indole in Tryptophan

During operations performed in acidic media tryptophan containing peptides
produce yellow, pink or purple colors due to several transformation products

of which here only the carboline derivatives

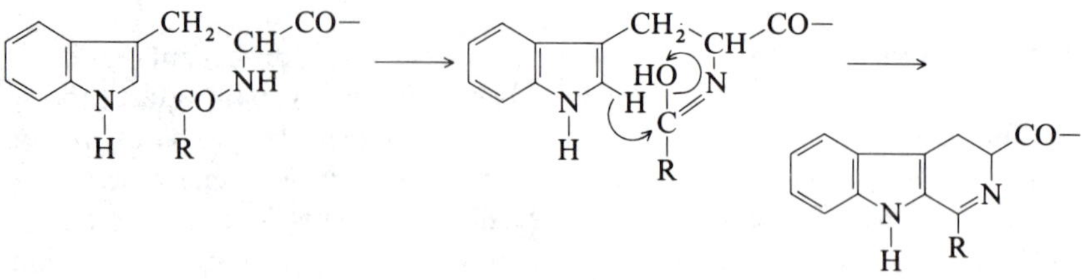

are mentioned. Low reaction temperature and exclusion of air greatly reduce the extent of decomposition.

## I. Side Reactions Caused by Steric Hindrance

In the discussion of side reactions we have dealt, so far, only with electronic effects. It would be a mistake, however, to underestimate the importance of geometric factors and, in particular, the influence of bulky side chains. Branching of the aliphatic chain in valine and in isoleucine is at the $\beta$-carbon atom, near to the reactive carbonyl group in activated intermediates. Hence, acylation reactions in which these amino acids are involved, do not proceed at a rate observed with other amino acids, for instance with leucine, in which branching is at the $\gamma$-carbon atom. Steric hindrance becomes even more pronounced when the activating group is similarly bulky as in pentachlorophenyl esters:

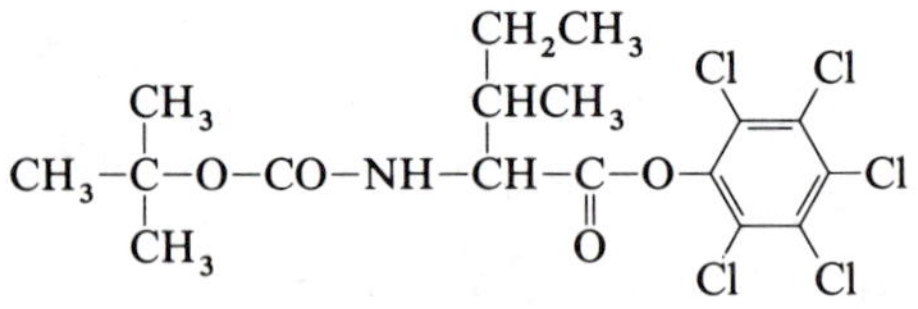

Further difficulties must be expected if the amine component contributes to an already existing steric effect. For instance, in the attempted preparation of prolyl-proline via a hydroxysuccinimide ester, attack on the activated ester carbonyl is accompanied by an attack on one of the succinimide carbonyl groups:

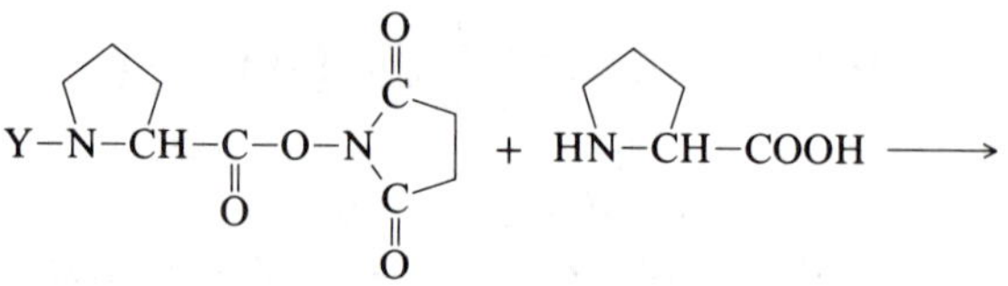

Y−N−CH−CO−N−CH−COOH +
60%

Y−N−CH−CO−O−NH−CO−CH$_2$CH$_2$−CO−N−CH−COOH
40%

Also, when the amino group to be acylated is somewhat hidden in the matrix of a polymeric support (cf. Chapter X), acylation might be incomplete even at equilibrium and the obstacle can not be overcome simply by raising the concentration of the acylating agent. Less bulky activating groups offer an escape from this dilemma.

We must add here that steric hindrance is not necessarily harmful. As mentioned before bulkiness in the activating portion of mixed anhydrides results in more unequivocal acylation reactions. Similarly, in $\beta$-tert.butyl esters of aspartyl residues ring-closure to aminosuccinyl peptides, observed with the corresponding $\beta$-benzyl esters, does not occur. Furthermore, formation of hydantoins and succinimide derivatives is considerably enhanced if glycine is involved in the process, because di-acylation of this unique amino acid without a side chain readily occurs. Accordingly, it is a mistake to use glycine in model experiments designed for the study of side reactions.


# J. Prevention and Suppression of Side Reactions


Because of the individuality of amino acids it is advisable to avoid generalization in peptide chemistry. Yet, from the foregoing and certainly not exhaustive discussion of side reactions we might conclude that most of them are caused by the presence of strong acids or excess base in the reaction mixture. Thus, schemes and conditions should be selected in which these harmful effects can be kept at a minimum. A second general measure, the use of the reactants in high concentration, follows from the intramolecular character of numerous side reactions. At high concentrations of both amine-component and acylating agent the desired reaction proceeds at better rate than in dilute solutions, hence competing side reactions become less significant. When poor solubility or high molecular weight of the amino-component prevents its use in high concentration, then the acylating agent, usually the derivative of a single amino acid, should be applied in an excess that provides for a concentration of at least $10^{-1}$ M throughout the coupling reaction. This measure (the "principle of excess") counteracts the decrease in rate which necessarily occurs in bimolecular reactions as the concentration of the reactants decreases. Therfore the extent of unimolecular side reactions, in which the rate is independent of concentration, can be markedly reduced. An excess on acylating agent also helps to achieve complete acylation of the amine-component and prevents thereby the formation of "deletion sequences", peptides from which one amino acid residue is missing.

A further concentration related problem has to be mentioned here. Following deprotection by acidolysis the regenerated amine is isolated as a salt of the acid used for cleavage. In the subsequent acylation step, however, the free amine is needed. Deprotonation with the help of ion exchangers can be applied or, in solid phase peptide synthesis (Chapter X) treatment with a tertiary amine and removal of the trialkylammonium salts by washing. In syntheses carried out in solution the general practice is to "liberate" the amine-component from its salt by adding an equimolar amount of a tertiary amine (triethylamine, diisopropylethylamine, N-methylmorpholine or N-ethylpiperidine) to the reaction mixture prior to coupling. It is obvious, however, that in many cases merely an equilibrium

$$R-NH_3^+ + R_3'N \rightleftharpoons R-NH_2 + R_3'NH^+$$

can be established. While it is true that during acylation of the amino group this equilibrium is shifted to the right, at any given time the concentration of the amine-component is lower than it would be if applied in completely deprotonated form. A notable exception is created by the insolubility of tertiary ammonium salts in certain solvents. For instance triethylamine hydrochloride, being practically insoluble in ethyl acetate, separates from the reaction mixture (if ethyl acetate is used as solvent for coupling) and shifts the equilibrium in the desired direction. The presence of tertiary amines is usually unfavorable during coupling: they can initiate side reactions through proton abstraction. These reactions can be suppressed by the addition of certain weak acids, for instance 2,4-dinitrophenol or pentachlorophenol, which show a distinct affinity for tertiary amines. They do not protonate the amine-component firmly enough to prevent its acylation. Moreover, the application of tertiary amines can be avoided by selecting highly acid sensitive amine-blocking groups and removing them with suitable weak acids, such as 1-hydroxybenzotriazole or tetrazole. The resulting salts are readily acylated, even with moderately reactive esters, without the addition of a tertiary base.

## Additional Source

Bodanszky, M., Martinez, J.: Side Reactions in Peptide Synthesis, in The Peptides, vol. 5, Gross, E., Meienhofer, J. eds., pp. 111–216, New York, Academic Press 1983

# VIII. Racemization

With the exception of glycine, in all amino acids that are constituents of proteins, the $\alpha$-carbon atom is chiral. In threonine and isoleucine a chiral center is present in the side chain as well. In order to secure the target peptide in homogeneous form it is absolutely essential to start from enantiomerically pure amino acids and to insist on conservation of chiral homogeneity throughout the various operations of synthesis. Otherwise, instead of a single product, a mixture of diastereoisomers will be obtained. Their number in a peptide with $n$ chiral centers is $2^n$. Accordingly, if racemization is not prevented, even in the synthesis of a moderately large peptide a complex mixture will be produced and separation of the desired material from a multitude of similar compounds might turn out to be an at least arduous and sometimes overwhelming task. Therefore, the importance of racemization studies and of the measures that must be taken for the prevention of any loss in chiral purity can not be overemphasized. In fact, "strategies" of peptide synthesis, that is general planning of schemes for syntheses (Chapter IX) are dictated primarily by considerations concerning conservation of chiral homogeneity.

## A. Mechanism of Racemization

With respect to chiral stability amino acids are fairly insensitive to acids and bases. Racemization via enolization of carboxylic acids in acidic solutions involves protonation of the carbonyl oxygen

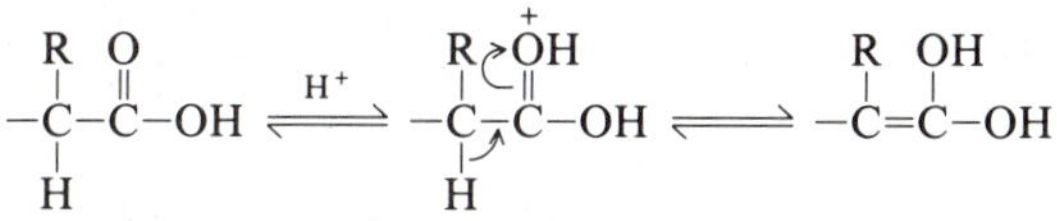

$$-\underset{\underset{H}{|}}{\overset{\overset{R}{|}}{C}}-\overset{\overset{O}{\|}}{C}-OH \;\underset{}{\overset{H^+}{\rightleftharpoons}}\; -\underset{\underset{H}{|}}{\overset{\overset{R}{|}}{C}}-\overset{\overset{+}{\overset{OH}{\|}}}{C}-OH \;\rightleftharpoons\; -\overset{\overset{R}{|}}{C}=\overset{\overset{OH}{|}}{C}-OH$$

yet, the presence of a nearby positively charged nitrogen atom hinders the formation of the second cation:

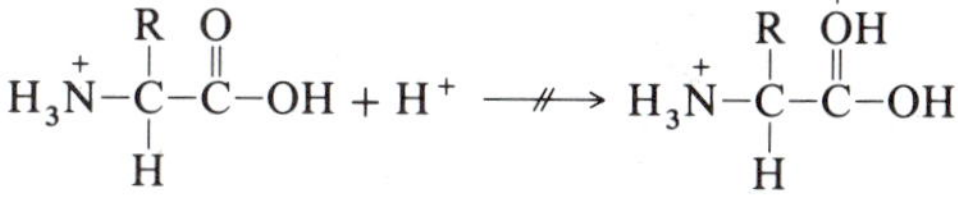

$$H_3\overset{+}{N}-\underset{\underset{H}{|}}{\overset{\overset{R}{|}}{C}}-\overset{\overset{O}{\|}}{C}-OH + H^+ \;\overset{}{\nrightarrow}\; H_3\overset{+}{N}-\underset{\underset{H}{|}}{\overset{\overset{R}{|}}{C}}-\overset{\overset{+}{\overset{OH}{\|}}}{C}-OH$$

The same can be said about base-catalyzed racemization of amino acids. A negative charge on the carboxylate hinders further proton abstraction from the α-carbon of amino acids; dianions

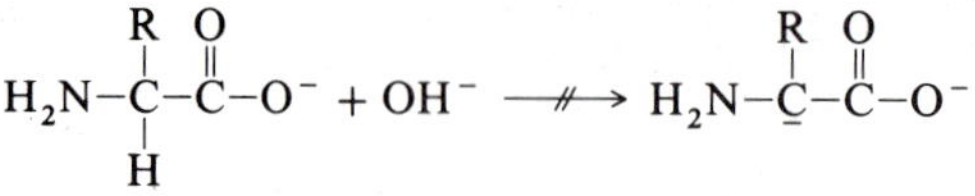

$$\text{H}_2\text{N}-\overset{\underset{|}{\text{H}}}{\underset{|}{\text{C}}}-\overset{\text{O}}{\underset{}{\text{C}}}-\text{O}^- + \text{OH}^- \longrightarrow\!\!\!\!\!/ \longrightarrow \text{H}_2\text{N}-\overset{\text{R}}{\underset{}{\text{C}}}-\overset{\text{O}}{\underset{}{\text{C}}}-\text{O}^-$$

are not stable in protic solvents. Therefore, operations involving the amino acids themselves, for instance introduction of the benzyloxycarbonyl group, are carried out in distinctly alkaline solution. In fact, excess alkali prevents the formation of reactive derivatives, such as mixed anhydrides, which might be prone to racemization. In the absence of a free carboxyl group, like in alkyl esters of peptides, base catalyzed racemization indeed does occur during saponification with alkali.

As indicated in the last paragraph, the *activated* carboxyl group poses the main problem in the preparation of optically homogeneous peptides. The electron-withdrawing effect of the activating group (x) extends to the α-carbon atom, the chiral center, and facilitates the abstraction of the hydrogen atom in the form of a proton

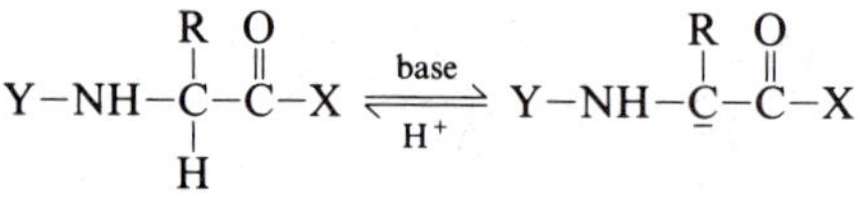

$$\text{Y}-\text{NH}-\overset{\underset{|}{\text{H}}}{\underset{|}{\text{C}}}-\overset{\text{O}}{\underset{}{\text{C}}}-\text{X} \underset{\text{H}^+}{\overset{\text{base}}{\rightleftharpoons}} \text{Y}-\text{NH}-\overset{\text{R}}{\underset{}{\text{C}}}-\overset{\text{O}}{\underset{}{\text{C}}}-\text{X}$$

This kind of simple proton abstraction is, however, not the sole and not even the most common mechanism of racemization. The most frequently invoked pathway involves cyclic intermediates, 4,5-dihydro-oxazole-5-ones or *azlactones:*

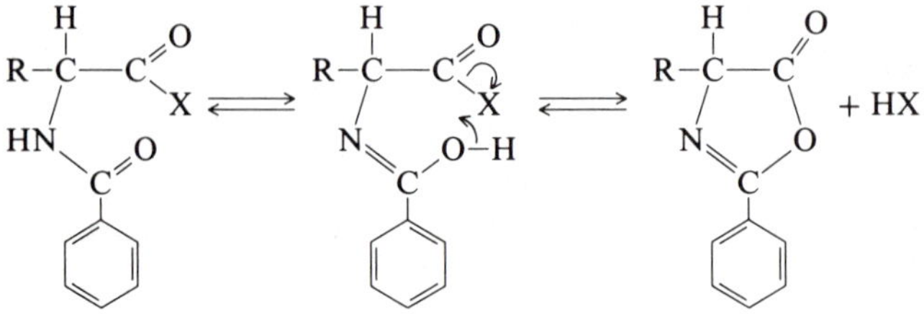

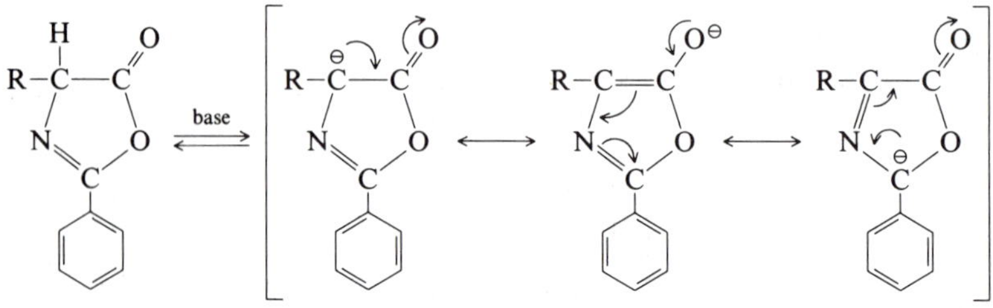

Proton abstraction from the chiral center yields a resonance stabilized carboanion that was first postulated and subsequently proven by ir spectra. Azlactone formation is quite pronounced in benzoylamino acids, less prevalent in acetylamino acids and was for a long time thought to be absent in amino acids acylated by benzyloxycarbonyl or other urethane type blocking groups. The absence of racemization on activation of the latter was attributed to lack of azlactone formation, but in recent years azlactones were obtained from benzyloxycarbonyl-, tert.butyloxycarbonyl etc. amino acids as well. Thus, azlactone formation itself is not a sufficient explanation of racemization; the stability of the cyclic intermediate toward bases must also be taken into consideration.

Activated derivatives of S-alkyl-cysteine suffer base catalyzed racemization even when their amino group is blocked by the benzyloxycarbonyl or other urethane-type protecting group. A simple, but not uncontested, explanation is reversible $\beta$-elimination

$$
\begin{array}{ccc}
\text{S–CH}_2\text{–C}_6\text{H}_5 & & \text{S–CH}_2\text{–C}_6\text{H}_5 \\
| & \xrightleftharpoons[\text{base}]{} & | \\
\text{CH}_2 & & \text{CH}_2 \\
| & & | \\
\text{Y–NH–CH–C–X} & & \text{Y–NH–C–C–X} \rightleftharpoons \\
\quad\quad\| & & \quad\quad\| \\
\quad\quad\text{O} & & \quad\quad\text{O}
\end{array}
$$

$$
\begin{array}{c}
\quad\quad\text{CH}_2 \\
\quad\quad\| \\
\text{Y–NH–C–C–X} \quad + \quad {}^-\text{S–CH}_2\text{–C}_6\text{H}_5 \\
\quad\quad\quad\| \\
\quad\quad\quad\text{O}
\end{array}
$$

supported by the isolation of the thiobenzyl ester of N-benzyloxycarbonyl-S-benzyl-DL-cysteine. This indicates that benzylmercaptane, one of the products of $\beta$-elimination, was present in the reaction mixture

$$
\begin{array}{l}
\quad\quad\quad\text{CH}_2\text{–C}_6\text{H}_5 \\
\quad\quad\quad| \\
\quad\quad\quad\text{S} \\
\quad\quad\quad| \\
\quad\quad\quad\text{CH}_2 \\
\quad\quad\quad| \\
\text{C}_6\text{H}_5\text{–CH}_2\text{O–CO–NH–CH–C–O–C}_6\text{H}_4\text{–NO}_2 \quad + \quad {}^-\text{S–CH}_2\text{–C}_6\text{H}_5 \longrightarrow \\
\quad\quad\quad\quad\quad\quad\quad\quad\quad\quad\| \\
\quad\quad\quad\quad\quad\quad\quad\quad\quad\quad\text{O}
\end{array}
$$

$$
\begin{array}{l}
\quad\quad\quad\text{CH}_2\text{–C}_6\text{H}_5 \\
\quad\quad\quad| \\
\quad\quad\quad\text{S} \\
\quad\quad\quad| \\
\quad\quad\quad\text{CH}_2 \\
\quad\quad\quad| \\
\text{C}_6\text{H}_5\text{–CH}_2\text{O–CO–NH–CH–CO–S–CH}_2\text{–C}_6\text{H}_5 \quad + \quad {}^-\text{O–C}_6\text{H}_4\text{–NO}_2
\end{array}
$$

Experiments with $S^{35}$ labeled benzylmercaptane, however, showed no incorporation of radioactivity. Also, racemization appears to be faster than deuterium exchange at the $\alpha$-carbon atom. Thus racemization via $\beta$-elimination might occur only at elevated temperature, while other mechanism(s) could be operative under the conditions usually maintained during coupling. Among the various hypotheses that were put forward the partial acceptance of the negative charge of the carbanion intermediate by the dorbitals of the sulfur atom is contradicted by the racemization of O-benzyl-serine derivatives.

The mechanisms described in the preceding paragraphs are the ones generally proposed for the explanation of racemization, but it is far from certain that other pathways are not involved. For instance it seems to be possible that the repeatedly observed loss of chiral integrity of the activated residue in coupling of peptides with the aid of dicyclohexylcarbodiimide is due to *intramolecular* proton abstraction by the basic center in the reactive O-acylisourea intermediate:

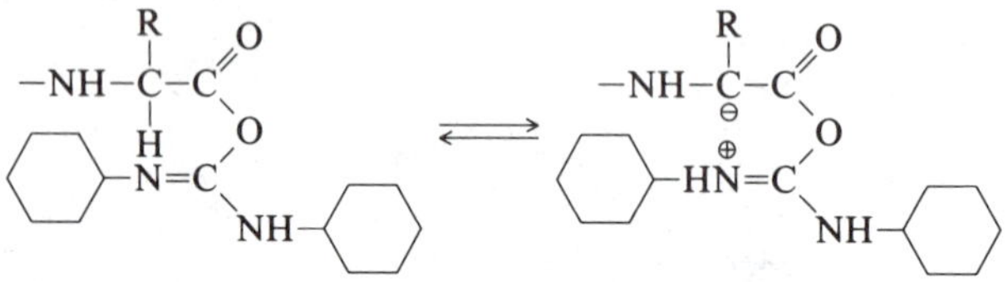

Thus the acidic character of additives such as 1-hydroxybenzotriazole contributes to their ability to prevent racemization in coupling with carbodiimides.

# B. Detection of Racemization

Loss of chiral homogeneity is an always present risk in peptide synthesis and there is an obvious need for methods that can reveal the presence of undesired diastereoisomers in the intermediates and particularly in the final product of the chainbuilding procedure. With carefully developed chromatographic systems it is often possible to separate fairly long peptide chains which are different from each other only with respect of the configuration of a singe amino acid residue. There are however several methods available for this kind of scrutiny that can be applied without a special study of the particular product in question. Such general methods require hydrolysis of the peptide either with *acid* or with the aid of *proteolytic enzymes*. The specificity of these enzymes is the major advantage of the enzymatic approach: no hydrolysis occurs between a D-residue and the next amino acid in the sequence. Therefore complete hydrolysis will take place only in peptides that contain no D residues. The rate of peptide bond fission, however, is a function of the amino acid cleaved from the N-terminus. Very slow rates can be achieved in the hydrolysis of glycyl and prolyl peptides with leucineaminopeptidase (a misnomer, since it is not specific for leucine), less difficulties are encountered with aminopeptidase M. Fast and complete hydrolysis of proline containing peptides requires the use of prolidases. A mixture of two enzymes, e.g. aminopeptidase M and prolidase can be quite efficient. Similar results are obtained with carboxypeptidases that provide stepwise removal of single amino acids starting with the C-terminal residue. Carboxypeptidase A has reduced catalytic effect when basic amino acids occupy the terminal position while carboxypeptidase B is most efficient just in this case. The yeast enzyme, carboxypeptidase Y is a more general catalyst.

A considerable number of biologically active peptides end with carboxamide rather than with a free carboxyl group. These peptide amides are, of course, no substrates for carboxypeptidases. An analogous problem exists in peptides which carry an acyl group such as the acetyl group at their N-terminus and, accordingly, can not be degraded with aminopeptidases. If both exopeptidase enzyme types fail one can resort to a preliminary fragmentation of the chain with endopeptidases, such as trypsin. The latter is very specific for basic amino acids and catalyzes hydrolytic cleavage of the bond between arginine and the next residue and of the bond that follows lysine. The tryptic fragments then are suitable for further enzymatic degradation with exopeptidases, particularly with carboxypeptidase B. Chymotrypsin is similarly useful, but its specificity is somewhat less pronounced: in addition to the bond which follows an aromatic amino acid, the bond after a leucine residue is also cleaved, albeit at a slow rate. In hydrolyzates obtained with proteolytic enzymes only amino acids should be present; uncleaved peptides reveal the presence of a D-residue.

Acid catalyzed hydrolysis followed by the identification of D-amino acids in the hydrolysate is equally useful. To make this possible the amino acids in the mixture are acylated with an enantiomerically pure amino acid, for instance with the N-carboxyanhydride of L-leucine. In the resulting mixture of dipeptides any racemized residue is revealed by the formation of *two* dipeptides that are diastereoisomers of each other, for instance L-leucyl-L-phenylalanine and L-leucyl-D-phenylalanine. Since these are compounds with different physical properties they are separable and appear as a doublet on recordings of an amino acid analzyer. In recent years the conversion to diastereoisomers became unnecessary because the availability of chiral supports now permits separation of enantiomers by high pressure liquid chromatography (HPLC) and also by thin layer chromatography on plates covered with a chiral layer.

# C. Racemization Studies in Model Systems

Racemization during the synthesis of peptides is a complex problem. The diversity of possible courses followed in the process is compounded by the individuality of the amino acids. This was already shown on the example of S-alkyl-cysteine residues which lose chiral purity by a special mechanism even if protected by a urethane-type protecting group that prevents racemization in other acylamino acids. The opposite end of the scale is represented by proline which, at least under the commonly applied conditions of peptide synthesis, resists racemization. This was conventionally explained by the circumstance that proline is a secondary amine and, therefore, in its N-acyl derivtives lacks the hydrogen atom which participates in the formation of azlactones (cf. section A in this chapter). The experience, however, gained with readily racemized N-acyl derivatives of N-methylamino acids contradicts this assumption. It appears more likely that the rigidity of the cyclic side chain of proline excludes certain transition states that are integral parts of the racemization process.

Various side chains affect the extent of racemization in different ways. Thus, the benzyl side chain in phenylalanine contributes to the stabilization of a carbanion and can thereby facilitate proton abstraction from the α-carbon atom. This effect is much more pronounced in phenylglycine (which is not a protein constituent but occurs in microbial peptides) because its chiral carbon atom is benzylic:

$$-NH-\underset{\underset{CH_2}{|}}{C}-CO- \qquad\qquad -NH-\underset{\underset{\phantom{CH_2}}{|}}{C}-CO-$$

The aliphatic side chains in alanine and leucine have no major influence but branching at the β-carbon atom in valine and isoleucine can enhance racemization because the combination of electron release and steric hindrance results in reduced coupling rates. The ensuing increase in the life-time of the reactive intermediate provides an extended opportunity for proton abstraction by base. It is obvious from these examples that the effect of individual side chains, the influence of various methods of coupling and the conditions of the peptide bond forming reaction (solvents, concentration, temperature, additives) must be studied in well designed experiments. Several model systems have been proposed for this purpose.

The first model (Williams and Young 1963) was based on coupling of benzoyl-L-leucine to glycine ethyl ester. The specific rotation of the crude product was used

$$\text{⟨⟩}-CO-NH-\underset{\underset{\underset{CH_2}{|}}{CH(CH_3)_2}}{\underset{|}{CH}}-\underset{\underset{O}{\|}}{C}-X + H_2N-CH_2-CO-OC_2H_5 \xrightarrow{(-HX)}$$

$$\text{⟨⟩}-CO-NH-\underset{\underset{\underset{CH_2}{|}}{CH(CH_3)_2}}{\underset{|}{CH}}-CO-NH-CH_2-CO-OC_2H_5$$

to establish enantiomeric purity. This simple system soon became popular and provided valuable information. Some shortcomings of the method must also be taken into consideration. The benzoyl group is not the best representative of blocking groups or of the part of the peptide chain that acylates the activated residue: it is more conducive to azlactone formation and might contribute to the stability of the anion generated from the azlactone by proton abstraction:

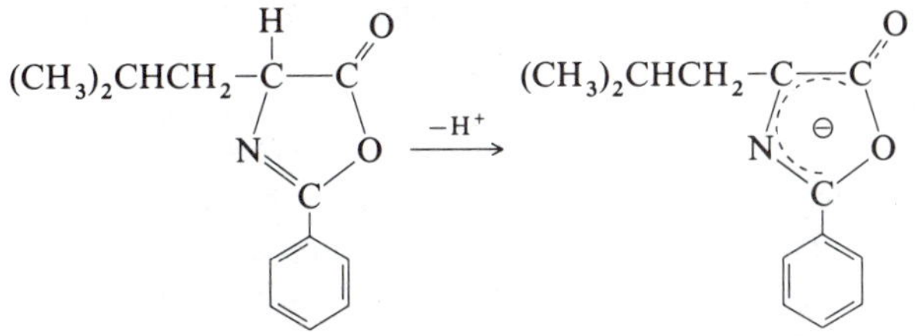

Therefore the Young-test might lead to somewhat exaggerated estimates of racemization. This distortion is counterbalanced by the relative resistance of leucine to racemization, but an additional problem is created by the necessity to isolate the crude benzoyl-leucyl-glycine ethyl ester in excellent yield. If less than near quantitative yield is achieved in coupling or in recovery of the product then it remains possible that the unaccounted portion contains a not insignificant amount of the D-isomer.

A frequently used early model (Anderson and Callahan 1958) is based on the coupling of benzyloxycarbonyl-glycyl-L-phenylalanine to glycine ethyl ester. Since the phenylalanine residue is acylated by glycine and not by the benzyloxycarbonyl group, it is not protected against racemization. Accordingly, reactions which cause loss of chiral purity produce in addition to Z-Gly-L-Phe-Gly-OEt also its enantiomer Z-Gly-D-Phe-Gly-OEt. The extent of racemization is easily established from the amount of the racemate because it separates from dilute ethanol. However, the results of this test are reliable only if the peptide bond forming reaction proceeds with excellent yield. The presence of byproducts can grossly interfere with crystallization and no racemate might separate although the D-isomer has been produced in considerable amount. In general: it is risky to rely on negative evidence, the *lack* of separation of the racemate.

Several later model systems were designed with the thought of separating products that are not enantiomers but diastereoisomers of each other. For instance in the coupling of acetyl-L-leucine to glycine ethyl ester racemization will yield acetyl-D-alloisoleucyl-glycine ethyl ester, because inversion at the α-carbon atom leads to a D-amino acid but chirality at the second chiral center, the β-carbon atom is unaffected

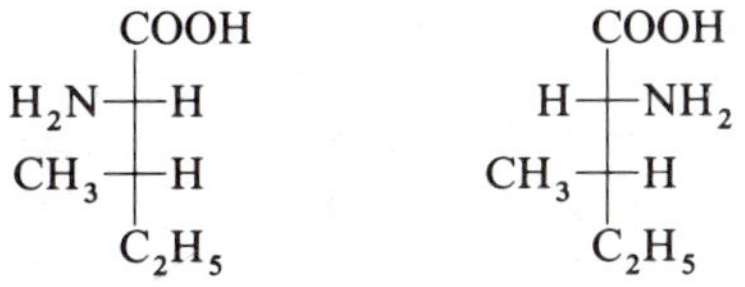

L-isoleucine          D-allo-isoleucine
(Fischer projections)

and hence, an alloisoleucine derivative is obtained. Complete hydrolysis (e.g. with 6 N HCl at 110° for 16 hrs) cleaves the amide and ester bonds and the hydrolysate can be applied to the column of an amino acid analyzer. In the well established Stein-Moore method of amino acid analysis isoleucine and alloisoleucine appear as well separated peaks and their ratio provides the sought information, the extent of racemization. The method does not require separation of the two peptides and therefore the results are not modified by imperfections in the operations of recovery. In a more general version of the same idea diastereomeric tripeptides are produced, deblocked and compared with the help of the amino acid analyzer as such, that is without hydrolysis. For instance benzyloxycarbonyl-glycyl-L-alanine is coupled to L-leucine benzyl ester and after hydrogenation the mixture containing glycyl-L-alanyl-L-leucine and glycyl-D-alanyl-L-leucine is applied to the column of the instrument. By replacing L-alanine with other L-amino acids important information can be gained about the sensitivity of various amino acids to a certain coupling method or the conditions of coupling.

Volatile peptide derivatives, for instance trifluoroacetyl-L-valyl-L-valine methyl ester or benzyloxycarbonyl-L-leucyl-L-phenylalanyl-L-valine tert.butyl ester can be separated from their diastereoisomers that contain a D-residue by vapor phase chromatography. Also, through the examination of nmr spectra of relatively simple peptides the extent of racemization that occured during their preparation can be determined without separation of the diastereoisomers, because the difference in the chemical shifts of some selected resonances is sufficient for integration. Thus the areas under the well separated peaks of the alanine methyl protons in acetyl-L-phenylalanyl-L-alanine methyl ester and in acetyl-D-phenylalanyl-L-alanine methyl ester can be integrated and the values used to determine the extent of racemization of the phenylalanine residue during coupling.

These are only selected examples of the numerous model systems proposed for the study of racemization, yet, even in such a brief treatment an approach based on enantio-selective enzymes should be mentioned. Coupling of benzyloxycarbonyl-L-alanyl-D-alanine to L-alanyl-L-alanine benzyl ester yields a blocked intermediate from which on catalytic hydrogenation the free L-Ala-D-Ala-L-Ala-L-Ala is obtained. This compound is completely resistant to hydrolysis catalyzed by aminopeptidases, because the first bond to be cleaved links the N-terminal residue to a D-amino acid. If however, racemization took place during coupling, this changed the activated residue, D-alanine to L-alanine and after deblocking the tetrapeptide L-Ala-L-Ala-L-Ala-L-Ala is obtained. The latter is completely digestible with aminopeptidases. The liberated alanine is determined and it is a rather exact measure of racemization because for each residue inverted four molecules of alanine are found in the analysis.

At this point a comment has to be added concerning the degree of racemization established with the help of model systems. Usually the amount of the undesired diastereoisomer is considered to represent the extent of racemization.

While this might be acceptable for the purpose of comparisons one has to keep in mind that from the achiral intermediate of the process the two isomers formed in equal amounts. Thus the number of molecules involved is twice the number of the undesired diastereoisomers formed.

# D. Prevention of Racemization

Since racemization during coupling is a base catalyzed process it is reasonable to assume that the nature of the base is not without influence on its outcome. Steric hindrance in some tertiary amines can weaken their ability to approach the chiral center in reactive intermediates. Diisopropylethylamine caused less racemization than triethylamine in the coupling of S-benzyl-L-cysteine derivatives, but it was without significant beneficial effect in reactions involving other amino acids. Perhaps in azlactones, the cyclic intermediates of racemization, the chiral carbon atom is well exposed and therefore the bulky groups in the tertiary amine can not interfere with proton abstraction. Also, some differences were found in couplings via mixed anhydrides between the previously preferred base triethylamine and tertiary amines such as N-ethylpiperidine or N-ethyl-morpholine, the latter being less conducive to racemization. The principal lesson to be learned is, however, to omit, when possible tertiary amines from the coupling mixture. The often applied approach, addition of a tertiary amine to a salt of the amine component is certainly inferior to the use of the amine component as such, that is the free amine. Several studies demonstrated that very little if any racemization takes place if this simple principle is followed.

Tertiary amines are added to the reaction mixture also when mixed anhydrides are generated:

$$R-COOH + NR' \longrightarrow R-COO^- \cdot H\overset{+}{N}R'_3$$

$$R-COO^- \cdot H\overset{+}{N}R_3 + R''-\overset{\overset{O}{\|}}{C}-Cl \longrightarrow R-\overset{\overset{O}{\|}}{C}-O-\overset{\overset{O}{\|}}{C}-R'' + R'_3\overset{+}{N}H \cdot Cl^-$$

Understandably, production of the same mixed anhydrides is accompanied by less racemization if it is carried out with the help of 1-ethyloxycarbonyl-2-eth-oxy-1,2-dihydroquinoline (EEDQ), because no addition of tertiary base is required and the basicity of the quinoline formed in the reaction is negligible. It is more difficult to counteract the effect of an intramolecular basic center, even if weak, such as the imidazole nucleus in the histidine side chain. While the here shown cyclization and enolization do not eliminate the ability of the activated species to acylate the amine component (acylimidazoles are good acylating agents) the chiral integrity of the histidine residue may certainly suffer in the process. Racemization via enolization might occur without cyclization as well, particularly because the enol can be stabilized in enolate form:

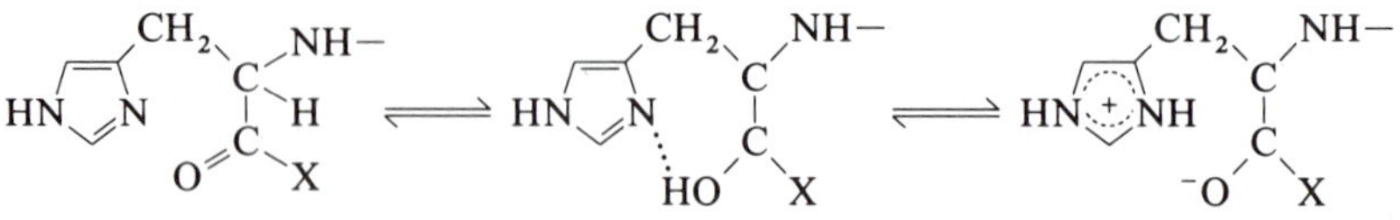

Hence it appears to be advantageous to further reduce the basic character of the imidazole by blocking, preferably at the $\pi$-nitrogen atom.

It is not surprising that a process that involves proton abstraction is influenced by the polarity of the solvent. Base catalyzed racemization of active esters is fast in polar solvents such as dimethylformamide and slow in non-polar media, for instance in toluene. It is rather unfortunate that such non-polar solvents are more often than not impractical in peptide synthesis. The poor solubility of most blocked intermediates in the commonly used organic solvents severely limits their use and in the preparation of larger peptides indeed dimethylformamide is most frequently applied. The problem of solubility is less serious in solid phase peptide synthesis (cf. Chapter X), where no real solvent is needed but merely a medium in which the polymeric support properly swells. This function is fulfilled by dichloromethane; its effect on racemization lies between the extremes mentioned.

Proton abstraction from the chiral carbon atom can be suppressed by the addition of weakly acidic materials to the reaction mixture. From the numerous additives tested 1-hydroxybenzotriazole (König and Geiger 1970a) and N-hydroxysuccinimide (Weygand et al. 1966) are routinely used in the practical execution of coupling. These compounds are not acidic enough to protonate the amino group

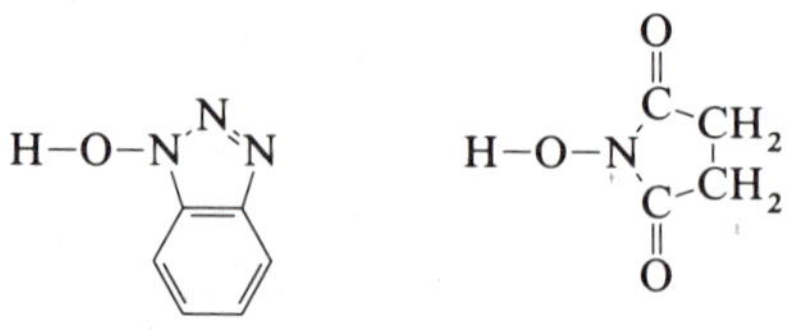

of the amine component and therefore they do not interfere with its acylation, but their acidity is sufficient to provide competition in abstraction of the proton from the carbon atom of activated intermediates. The significance of these additives is based however, not merely on their acidic character: many other weak acids perform poorly in the role of racemization suppressing agents. Both additives are related to hydroxylamine and function as powerful auxiliary nucleophiles. They react with overactivated intermediates, such as the O-acylisourea in carbodiimide mediated couplings

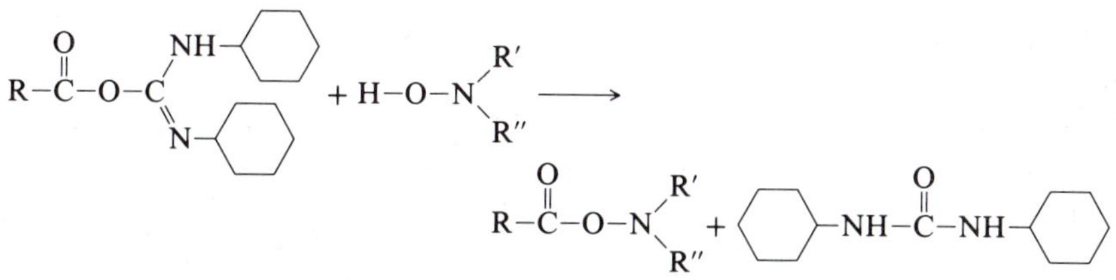

reducing thereby the lifetime of the racemization prone species. The active esters produced in these reactions have higher chiral stability. In their reaction with the amine-component the additive is regenerated and assures a continued beneficial effect. Strangely, the highly efficient additives 3-hydroxy-3,4-dihydro-1,2,3-benzotriazine-4-one (König and Geiger 1970b) and 2-hydroximinocyanoacetic acid ethyl ester (Itoh 1973) have not been widely used so far although their effect on the prevention of racemization exceeds that of the popular 1-hydroxybenzotriazole.

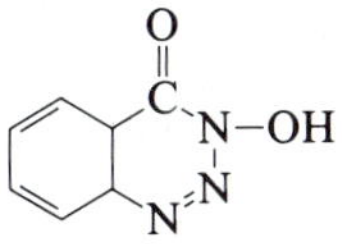

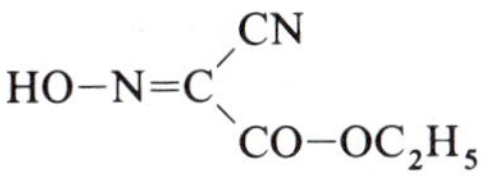

The rather general measure that can be taken against side reactions, the use of both the carboxyl-component and the amine-component in high concentration, is applicable for the suppression of racemization as well. However, poor solubility of intermediates, sometimes even in dimethylformamide, presents a formidable obstacle compounded by the high molecular weight of some blocked peptides. Thus a high molar concentration of both components is often unattainable. On the other hand if the carboxyl component is not a peptide derivative but rather the blocked and activated form of a single amino acid, it can be used in excess. This excess can be adjusted to provide for a concentration which remains sufficiently high (for instance more than 0.1 molar) throughout the coupling reaction. The relative simplicity of the blocked *and* activated amino acid derivatives and their availability from commercial sources render this sacrifice usually acceptable. Over and above the possibility of performing coupling reactions according to the *principle of excess,* addition of single amino acid residues has a further important benefit in the conservation of chiral purity: the most commonly used amine-protecting groups, the benzyloxycarbonyl and the tert.butyloxycarbonyl group, efficiently prevent racemization in most cases. As mentioned before, this is a common feature of urethane-type blocking groups and applies for the acid-stable, base-sensitive 9-fluorenylmethyloxycarbonyl (Fmoc) group as well.

In the preceding discussion we have dealt only with racemization during peptide bond formation. Loss of chiral purity can occur, however, also during certain processes of deprotection. Hydrogenolysis is quite innocuous in this respect and acidolysis is harmful only if it is carried out under drastic conditions, such as elevated temperature. Saponification of esters with alkali can cause measurable racemization. This must be kept at a minimum by carrying out the reaction at ice-bath temperature, preferably at constant $p_H$. Large excess of alkali certainly must be avoided. The presence of $Cu^{++}$ ions prevents racemization in alkaline hydrolysis (and probably also in many instances of coupling), but complete removal of the metal ions from the complex is not always straightforward.

# References

Anderson, G. W., Callahan, F. M.: J. Amer. Chem. Soc. *80,* 2902 (1958)
Itoh, M.: Bull. Chem. Soc. Jpn. *46,* 2219 (1973)
König, W., Geiger, R., a: Chem. Ber. *103,* 788 (1970); b: Chem. Ber. *103,* 2034 (1970)
Weygand, F., Hoffmann, D., Wünsch, E.: Z. Naturforsch. *21b,* 426 (1966); cf. also Wünsch, E., Drees, F.: Chem. Ber. *99,* 110 (1966)
Williams, M. W., Young, G. T.: J. Chem. Soc. 881 (1963); cf. also Smart, N. A., Young, G. T., Williams, M. W.: ibid. p. 3902

# Additional Sources

Kemp, D. S.: Racemization in Peptide Synthesis in The Peptides, vol. 1, Gross, E., Meienhofer, J., eds., pp. 315–383, New York, Academic Press 1979

# IX. Design of Schemes for Peptide Synthesis

In the strategical planning that must precede the synthesis of a larger peptide racemization is one of the most important considerations. Therefore, it seems to be appropriate to discuss the various schemes of synthesis at this point. Due to the individuality of amino acid residues and to variations in the properties of blocked intermediates it appears to be impractical to propose a general scheme (strategy) that would be applicable for any peptide. Peptide synthesis should be based on retrosynthetic analysis, starting with identification of the problems inherent in the sequence of the target compound.

In principle three approaches are possible: A. condensation of peptide segments, B. stepwise chain-building starting with the N-terminal residue and C. stepwise chain building starting at the C-terminus. We will attempt to evaluate these alternatives, but with some reservation: there is no consensus among peptide chemists in this area.

## A. Segment Condensation

In the earliest days of practical peptide synthesis, in the preparation of the nonapeptide oxytocin or the octapeptide angiotensin, segment condensation appeared to be the obvious strategy. Reduction of a major task to smaller problems, a Cartesian approach, is clearly attractive. Equally important is, however, the possibility of dividing the effort between members of a team. Preparation of the individual segments, often dipeptides, could be entrusted to less experienced coworkers while the arduous task of segment condensation to an adept in peptide chemistry. A similar distribution of responsibilities is not feasible in stepwise chain lengthening. These considerations must have guided the investigators who undertook the synthesis of biologically active peptides in the nineteen fifties. The retrosynthetic scheme for the synthesis of the octapeptide angiotensin is shown here as an example:

Asp – Arg – Val – Tyr – Ile – His – Pro – Phe
⇩
Asp – Arg – Val – Tyr + Ile – His – Pro – Phe
⇩                      ⇩
Asp – Arg + Val – Tyr   Ile – His + Pro – Phe
⇩      ⇩        ⇩       ⇩
Asp + Arg   Val + Tyr   Ile + His   Pro + Phe

Four dipeptides are combined to give two tetrapeptides and construction of the chain is concluded with the condensation of the tetrapeptides. At this point we disregard the problems related to residues with functional side chains, the side reactions that can take place during the incorporation of aspartic acid or valine and also the by-products generated in the process of deprotection because we wish to concentrate our attention on the question of racemization. Assuming the use of urethane-type amine-protecting groups, no loss in chiral homogeneity should be expected in the synthesis of the four dipeptide units. In the condensation of Asp-Arg with Val-Tyr, however, racemization of the arginine residue might occur because the urethane-type blocking group in the dipeptide Asp-Arg is attached to Asp and not to the activated residue, Arg. Similarly, combination of the partially blocked derivatives of the dipeptides Ile-His and Pro-Phe could yield a tetrapeptide derivative, contaminated by the diastereoisomer in which a D-His residue is present instead of the desired L-His. Unless proper precautions are taken the same shortcoming will reoccur in the final condensation of the two tetrapeptide derivatives and some of a D-tyrosine containing analog of angiotensin will also be formed. In summary, no loss of chiral integrity took place in the incorporation of *single amino acid residues,* because these were provided with blocking groups that prevent racemization, but racemization could occur during the combination of *peptides* because in these the racemization preventing blocking groups are attached to the N-terminal residue of the activated peptide and not to the C-terminal residue that has the activated carboxyl. With the advent of additives such as 1-hydroxybenzotriazole, N-hydroxysuccinimide, 3-hydroxy-3,4-dihydro-1,2,3-benzotriazine-4-one or 2-hydroximinocyanoacetic acid ethyl ester, condensation of segments can be carried out with little racemization, but the problem is not entirely solved. Evidence supporting the racemization preventing effect of these additives was obtained in model experiments, in which the two components to be coupled were present in fairly high concentration. The situation can be substantially different in actual syntheses where segments of large molecular weight are coupled. Since it might be impractical to insist on high molar concentration in these cases the beneficial effect of the additives could be less than expected. A similar relationship exists between coupling of small versus combining of large segments also in respect to other side reactions. For instance the Curtius rearrangement of acid azides usually yields negligible amounts of urea derivatives in model experiments, but acylation of a larger amine-component with the azide of a high molecular weight carboxyl-component often gave disappointing results.

In adoption of the segment condensation strategy retrosynthetic analysis of the target peptide is a worthwhile undertaking. For instance, racemization can be completely avoided if segments having glycine for C-terminal residue are selected, and even with proline in this position practically no racemization will occur. Also, alanine or leucine are less prone to lose chiral purity than phenylalanine or histidine. Thus, a judicious dissection of the sequence into segments can greatly improve the results.

## B. Stepwise Chain-lengthening Starting with the C-terminal Residue

This approach appears to be attractive first and foremost because it is Nature's way to synthesize proteins on the ribosomes. Furthermore, as the mere inspection of the retrosynthetic scheme for angiotensin shows

Asp – Arg – Val – Tyr – Ile – His – Pro – Phe
⇩
Asp – Arg – Val – Tyr – Ile – His – Pro + Phe
⇩
Asp – Arg – Val – Tyr – Ile – His + Pro
⇩
Asp – Arg – Val – Tyr – Ile + His
⇩
Asp – Arg – Val – Tyr + Ile
⇩
Asp – Arg – Val + Tyr
⇩
Asp – Arg + Val
⇩
Asp + Arg

only a single residue, the N-terminal Asp requires an amine-blocking group which is removed after completion of chain-building. Yet, this advantage is also the greatest disadvantage of the approach: the blocking group protects the Asp residue against racemization in the synthesis of the dipeptide derivative Y-Asp-Arg but in subsequent steps blocked *peptides* are activated and their activated C-terminal residues are exposed to the racemizing effect of the coupling procedure without protection for chiral integrity. This circumstance is important enough to render the strategy impractical. It has been followed only rarely and the investigators who adopted this approach failed to demonstrate the chiral purity of their products.

## C. Stepwise Chain-lengthening Starting with the C-terminal Residue

This strategy would seem too laborious for practical purposes. It requires the incorporation of each residue in blocked and activated form and removal of the amine-blocking group after each chain lengthening step.

$$Asp - Arg - Val - Tyr - Ile - His - Pro - Phe$$
$$\Downarrow$$
$$Asp + Arg - Val - Tyr - Ile - His - Pro - Phe$$
$$\Downarrow$$
$$Arg + Val - Tyr - Ile - His - Pro - Phe$$
$$\Downarrow$$
$$Val + Tyr - Ile - His - Pro - Phe$$
$$\Downarrow$$
$$Tyr + Ile - His - Pro - Phe$$
$$\Downarrow$$
$$Ile + His - Pro - Phe$$
$$\Downarrow$$
$$His + Pro - Phe$$

$$Pro + Phe$$

The ability, however, of a whole class of protecting groups to prevent racemization of the residue to which they are attached renders this approach not only practical but probably also the most useful of the three alternatives. The efficiency of the *stepwise strategy* (Bodanszky 1960) is due in part to the application of acylating agents in high enough concentration to secure reasonably high reaction rates throughout the coupling (principle of excess). When active esters of blocked amino acids are used for the incorporation the excess reagent remains unchanged and can often be recovered.

Even from this brief discussion of strategies it is clear that there are only two schemes to chose from: segment condensation and stepwise chain lengthening from the C-terminus. The earliest syntheses of biologically active preptides followed the segment condensation approach and it is still practiced in numerous laboratories. The practicality of the stepwise strategy had first to be demonstrated for its acceptance and even the successful syntheses of oxytocin (Scheme 2) and secretin failed to convince the majority of the investigators about its advantages. The need to carry out long series of coupling and deblocking opertions almost singlehandedly was probably the main obstacle to the implementation of this approach. Yet, with the advent of techniques of facilitation, particularly with the development of solid phase peptide synthesis (cf. Chapter X) the stepwise strategy became fully appreciated and gained broad application. In major syntheses carried out in solution most investigators prefer to assemble moderate size peptides in stepwise manner and to combine these into larger segments and finally into the molecule of the target compound through coupling steps in which racemization is suppressed by the addition of the already discussed auxiliary nucleophiles. This measure is usually sufficient for the conservation of chiral purity but remains often unsatisfactory in terms of yields. With large molecular weight segments some of the problems caused by various side reactions can not be fully solved by the recommended high concentration of the reactants. An interesting compromise was reached in the synthesis of ribonuclease. The 124 residue chain of this enzyme was constructed (Yajima

**Scheme 2.** Stepwise synthesis of Oxytocin [a]

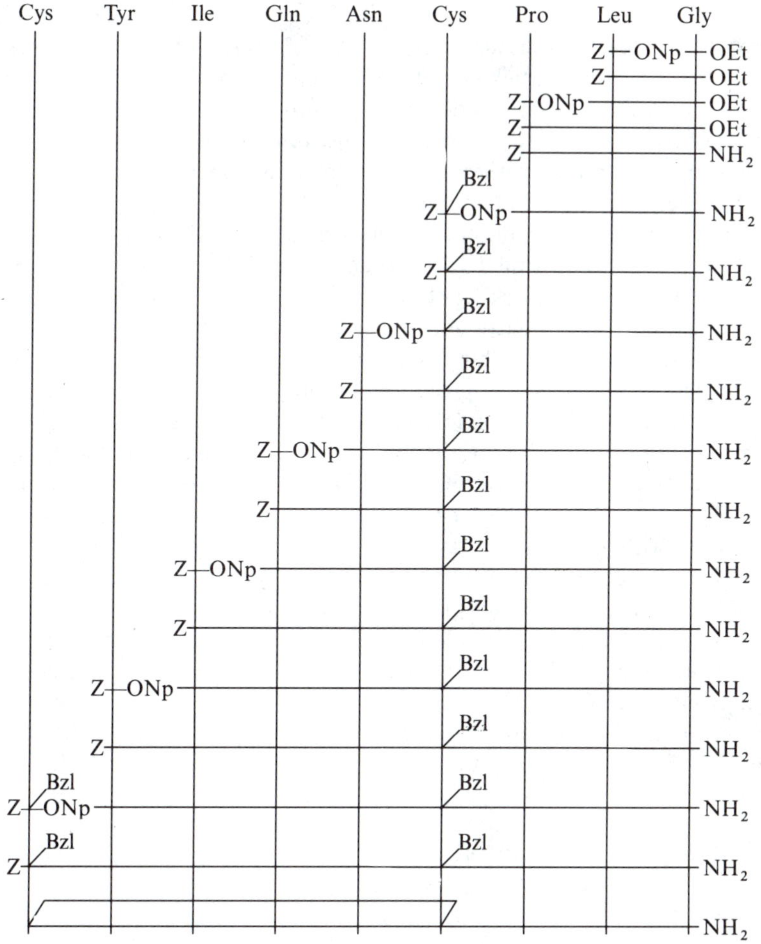

| Cys | Tyr | Ile | Gln | Asn | Cys | Pro | Leu | Gly |
|-----|-----|-----|-----|-----|-----|-----|-----|-----|
| | | | | | | Z—ONp—OEt | | |
| | | | | | | Z—OEt | | |
| | | | | | Z—ONp | OEt | | |
| | | | | | Z— | OEt | | |
| | | | | | Z— | NH$_2$ | | |
| | | | | Bzl | | | | |
| | | | | Z—ONp | NH$_2$ | | | |
| | | | | Bzl | | | | |
| | | | | Z— | NH$_2$ | | | |
| | | | Bzl | | | | | |
| | | | Z—ONp | NH$_2$ | | | | |
| | | | Bzl | | | | | |
| | | | Z— | NH$_2$ | | | | |
| | | Bzl | | | | | | |
| | | Z—ONp | NH$_2$ | | | | | |
| | | Bzl | | | | | | |
| | | Z— | NH$_2$ | | | | | |
| | Bzl | | | | | | | |
| | Z—ONp | NH$_2$ | | | | | | |
| | Bzl | | | | | | | |
| | Z— | NH$_2$ | | | | | | |
| Bzl | Bzl | | | | | | | |
| Z—ONp | | | NH$_2$ | | | | | |
| Bzl | Bzl | | | | | | | |
| Z— | | | NH$_2$ | | | | | |
| | | | | | | | | NH$_2$ |

[a] Bodanszky and du Vigneaud, 1959; ONp = *p*-nitrophenyl ester

and Fujii 1980) from small segments each containing only a few residues which then were assembled in a stepwise manner starting with the C-terminal portion of the chain. In order to reduce racemization the individual segments were activated in the form of their acid azides and applied *in excess*. In this way the difficulties caused by the poor solubility of large segments were at least in part circumvented and the extent of side reactions, such as Curtius rearrangement of the acid azides was greatly reduced.

# D. Tactical Considerations

Most of the side reactions discussed in Chapter VII occur during activation and coupling. Imperfections which stem from the operations of deprotection are, however, not less important and can be serious obstacles to achieving the synthesis of homogeneous target peptides. Because of the repetitive character of deblocking in stepwise synthesis the selection of methods for protection and unmasking requires special attention.

Acidolysis at two widely different levels of acidity, mostly with trifluoroacetic acid for the removal of the temporary masking of the α-amino groups and with liquid HF for the deblocking of the semipermanent blocking of side chain functions, is commonly applied. Yet, it is clear that the ratio of acidolysis rates for the two different kinds of blocking groups is a finite number and that the relatively acid stable side chain protecting groups are not completely resistant under the conditions of mild acidolysis. Hence, orthogonal combinations that rely on *two different mechanisms* should give more unequivocal results. For instance a choice of two sets of groups, one cleaved by hydrogenation, the other by acidolysis seems to be superior to the non-orthogonal combination of acid labile groups with different sensitivity toward acids. The orthogonal combination of the base labile 9-fluorenylmethyloxycarbonyl (Fmoc) group with acid labile side chain blocking groups based on the formation of the tert.butyl cation has already been mentioned but deserves further emphasis at this point.

An attractive possibility in the selection of blocking groups is to provide for simultaneous deprotection at the conclusion of the chain building procedure. For instance the final nonapeptide derivative in the first synthesis of the hormone oxytocin (du Vigneaud et al. 1953)

$$
\begin{array}{ccccccc}
\text{Bzl} & & & & \text{Bzl} & & \\
| & & & & | & & \\
\text{Z}-\text{Cys}-\text{Tyr}-\text{Ile}-\text{Gln}-\text{Asn}-\text{Cys}-\text{Pro}-\text{Leu}-\text{Gly}-\text{NH}_2
\end{array}
$$

was unmasked in a single operation by reduction with sodium in liquid ammonia as was the blocked nonapeptide with the sequence of lysine vasopressin

$$
\begin{array}{ccccccc}
\text{Bzl} & & & & \text{Bzl} & & \text{Tos} \\
| & & & & | & & | \\
\text{Z}-\text{Cys}-\text{Tyr}-\text{Phe}-\text{Gln}-\text{Asn}-\text{Cys}-\text{Pro}-\text{Lys}-\text{Gly}-\text{NH}_2
\end{array}
$$

In a similar vein the benzyloxycabonyl group was chosen for blocking the α-amino group of the N-terminal residue in the first synthesis of porcine secretin (Bodanszky et al. 1966), because in this way a single step, catalytic hydrogenation, could accomplish final deprotection:

$$
\begin{array}{c}
\text{Bzl} \quad \text{OBzl} \qquad\qquad\qquad \text{Bzl} \quad \text{OBzl} \qquad \text{Bzl} \quad \text{NO}_2 \qquad\quad \text{NO}_2 \\
| \qquad | \qquad\qquad\qquad\qquad\quad | \qquad | \qquad\quad | \qquad | \qquad\qquad | \\
\text{Z}-\text{His}-\text{Ser}-\text{Asp}-\text{Gly}-\text{Thr}-\text{Phe}-\text{Thr}-\text{Ser}-\text{Glu}-\text{Leu}-\text{Ser}-\text{Arg}-\text{Leu}-\text{Arg}-
\end{array}
$$

$$
\begin{array}{c}
\text{OBzl} \ \ \text{Bzl} \qquad \text{NO}_2 \qquad\quad \text{NO}_2 \\
| \qquad | \qquad\quad | \qquad\qquad | \\
\text{Asp}-\text{Ser}-\text{Ala}-\text{Arg}-\text{Leu}-\text{Gln}-\text{Arg}-\text{Leu}-\text{Leu}-\text{Gln}-\text{Gly}-\text{Leu}-\text{Val}-\text{NH}_2
\end{array}
$$

Special consideration should be given to the selection of temporary protection of the α-amino group of the N-terminal residue of peptides that serve as amino component in a subsequent step of segment condensation. For instance the base labile 9-fluorenylmethyloxycarbonyl group yields, after deprotection, a peptide with an unprotonated amino group in contrast with the amine salts formed in acidolysis. The advantage of unprotonated amines has already been discussed, thus it is clear that the proper selection of the blocking group at such crucial points has a major influence on the quality of the large intermediates and even of the final product.

# E. Synthesis of Special Target Peptides

So far we have been concerned mainly with methods leading to peptide chains, usually single chains, with a well defined sequence. There are however peptide chains of considerable length that contain only a single kind of amino acid. Such *polyamino acids* (or amino acid polymers) can be useful as models of proteins in physico-chemical studies or might have special roles, as for instance polylysine in the construction of antigens or polyleucine as chiral support in the chromatographic separation of enantiomers. Other long peptide chains are composed of more than one kind of amino acid, random copolymers or two (or more) amino acids in a sequence that is repeated many times, e.g. Ala-Gly-Pro-Ala-Gly-Pro-Ala-Gly-Pro-Ala-Gly-Pro-Ala-Gly-Pro . . . or Ala-Gly-Pro-(Ala-Gly-Pro-)$_n$-Ala-Gly-Pro where $n$ can be a large number. Synthesis of such "sequential polypeptides" was often pursued because they are useful in the study of skeletal proteins such as collagen. An at least equally important objective is the preparation of cyclic peptides. Several hormones, like the already mentioned oxytocin and vasopressins, somatostatin and others have ring structures and peptide antibiotics are cyclic. Hence cyclization of their open chain ("linear") peptide precursors is an important topic. Last but not least, peptides containing a non-amino acid or modified amino acid components are used with increasing frequency for biological and even medical purposes, for example as enzyme inhibitors. Problems related to the synthesis of such special peptides are the subject of the following sections.

## 1. Polyamino Acids and Sequential Polypeptides

High molecular weight materials form from several amino acids, particularly from glycine, at elevated temperatures. Also, complex, protein like substances can be produced from mixtures of amino acids by heating or with the help of various energy sources, such as ultraviolet radiation or bombardment with slow neutrons. These observations and materials gave rise to interesting speculations about the origin of life, but they seem to lie outside the framework of this volume that deals with well defined peptides.

Preparation of polyamino acids through N-carboxyanhydrides (NCA-s or Leuchs' anhydrides) is a classical approach. These compounds, oxazolidine-2,5-diones are readily obtained from amino acids on treatment with phosgene in aprotic solvents, followed by heating:

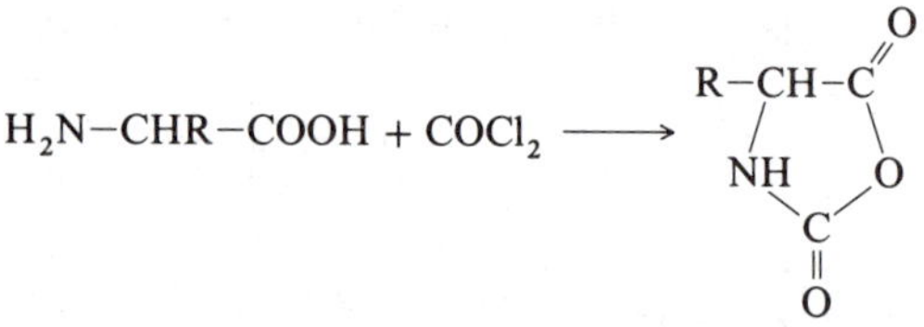

Alternatively they are prepared from alkoxycarbonylamino acid chlorides via thermal elimination of alkyl chloride:

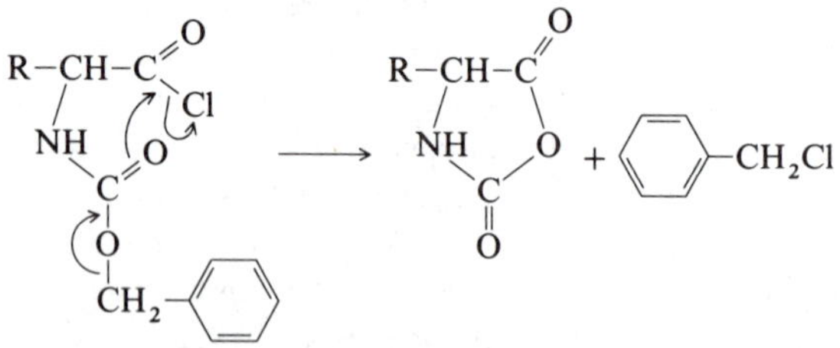

Polycondensation of highly reactive intermediates is usually described in a simple fashion. Ring opening by a trace of water yields a carbamoic acid which looses carbon dioxide and regenerates the amino acid:

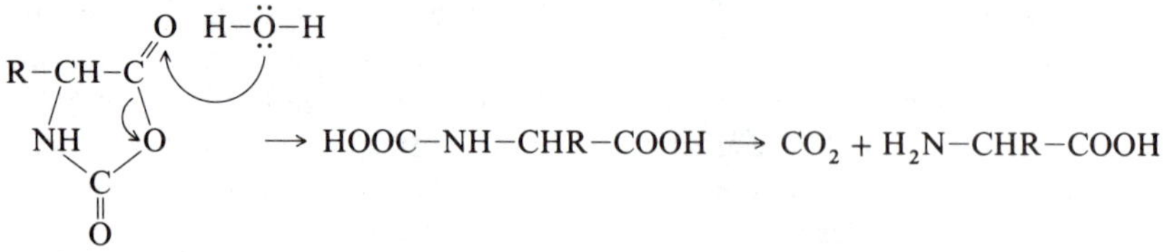

$$\rightarrow HOOC-NH-CHR-COOH \rightarrow CO_2 + H_2N-CHR-COOH$$

The amino group of the amino acid attacks the more reactive carbonyl group (the carboxyl-carbonyl) in a second molecule of the anhydride and the coupling reaction is followed by decarboxylation to yield a dipeptide:

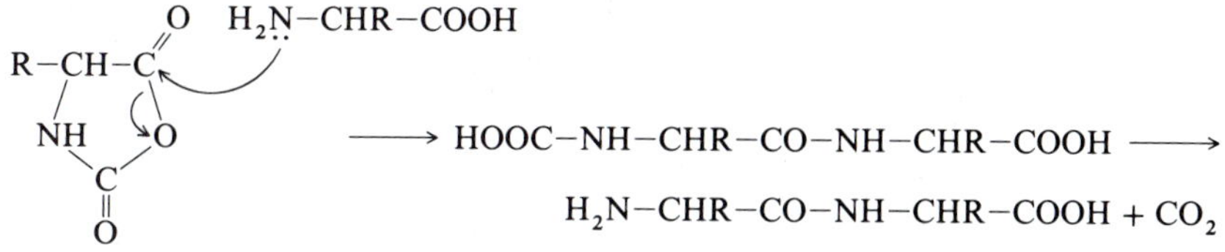

$$R-CH-C \overset{O}{\underset{NH \quad O}{\diagdown}} \quad H_2N-CHR-COOH$$

$$\longrightarrow \quad HOOC-NH-CHR-CO-NH-CHR-COOH \longrightarrow$$

$$H_2N-CHR-CO-NH-CHR-COOH + CO_2$$

The process continues through the attack of the amino group of the dipeptide on a further molecule of the anhydride

$$R-CH-C \quad + H_2N-CHR-CO-NH-CHR-COOH \longrightarrow$$

$$H_2N-CHR-CO-NH-CHR-CO-NH-CHR-COOH + CO_2$$

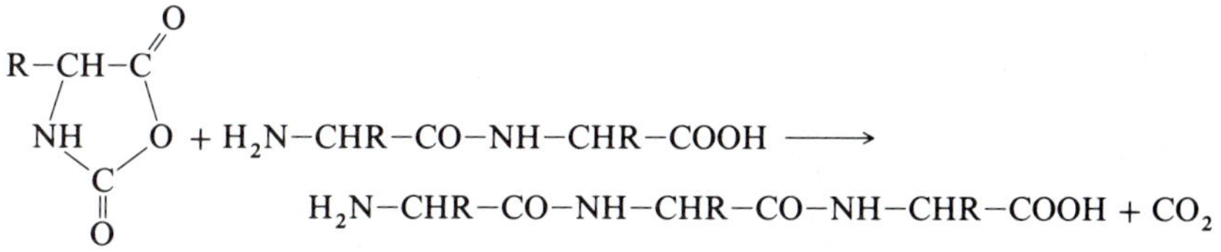

and high molecular weight materials form in this manner. In the actual execution of the polymerization however a mixture of chains of various lengths is produced. The molecular weight distribution of the product varies with conditions of the reaction and is not easily controlled. An additional complexity in the composition of the polymerization product arises from the attack of the nucleophile(s) on the less reactive carbonyl group, a minor but not negligible side reaction yielding substituted ureas

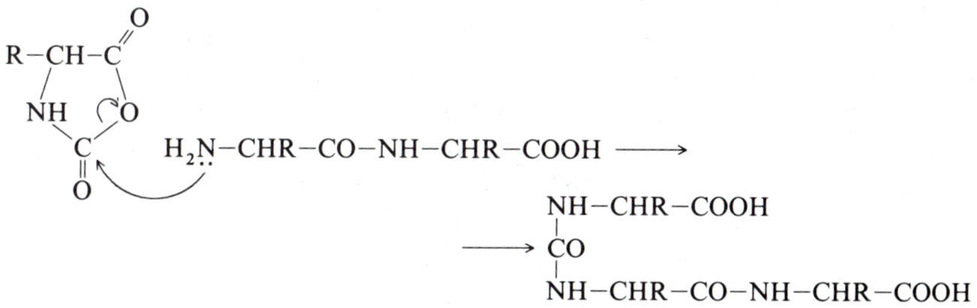

$$R-CH-C \quad H_2N-CHR-CO-NH-CHR-COOH \longrightarrow$$

$$\longrightarrow \quad \begin{array}{c} NH-CHR-COOH \\ | \\ CO \\ | \\ NH-CHR-CO-NH-CHR-COOH \end{array}$$

which render the synthetic molecules less similar to peptide chains in proteins. Understandably, better homogeneity can be expected, with respect both to molecular weight and peptide linkage in a less well known approach (Miyoshi et al. 1970) based on the systematic build-up of blocked and activated segments according to schemes such as the one shown here:

$$Pro + Pro \longrightarrow Pro - Pro$$

$$Pro - Pro + Pro - Pro \longrightarrow Pro - Pro - Pro - Pro$$

$$Pro - Pro - Pro - Pro + Pro - Pro - Pro - Pro \longrightarrow$$

$$Pro - Pro - Pro - Pro - Pro - Pro - Pro - Pro$$

Synthesis of sequential peptides usually relies on an activated derivative of the repeating unit from which the amine-protecting group is removed by acidolysis. Pentachlorophenyl esters were found to produce high molecular weight materials. The reaction is started by deprotonation with a tertiary amine but the addition of a small amount of a secondary amine as initiator is also advisable:

$$Z - Ala - Pro - Gly - OPcp \xrightarrow{HBr/AcOH} Ala - Pro - Gly - OPcp.HBr \xrightarrow{N(C_2H_5)_3}$$

$$Ala - Pro - Gly - OPcp \ (I); \ I + HNR_2 \longrightarrow Ala - Pro - Gly - NR_2 \xrightarrow{I}$$

$$Ala - Pro - Gly - Ala - Pro - Gly - NR_2 \xrightarrow{I} Ala - Pro - Gly - Ala - Pro - Gly -$$

$$Ala - Pro - Gly - NR_2 \xrightarrow{I} Ala - Pro - Gly - Ala - Pro - Gly - Ala - Pro - Gly -$$

$$Ala - Pro - Gly - NR_2$$

$$OPcp = -O-\overset{Cl \quad Cl}{\underset{Cl \quad Cl}{\bigcirc}}-Cl \quad R = C_2H_5$$

A sequential polymer of well defined molecular weight was assembled (Sakakibara et al. 1968) on an isoluble polymeric support by the repeated incorporation of the tripeptide sequence Pro-Pro-Gly. Twenty repetitions afforded a chain of 60 residues and the physico-chemical properties of this material were superior to those obtained by polymerization of an active ester of the tripeptide. In an alternative approach (Kemp et al. 1974) the sequence Gly-Leu-Gly was doubled by the reaction of an activated derivative of Z-Gly-Leu-Gly with the free tripeptide, followed by the removal of the blocking group from part of the product. Activation of the remaining portion of the hexapeptide derivative Z-Gly-Leu-Gly-Gly-Leu-Gly and reaction of the activated derivative with a suspension of the free hexapeptide in dimethylsulfoxide gave the partially blocked 12-peptide Z-Gly-Leu-Gly-Gly-Leu-Gly-Gly-Leu-Gly-Gly-Leu-Gly and through repeated doublings a 24-peptide and finally a sequential polymer containing 48 residues could be secured.

## 2. Disulfide Bridges

Intramolecular reaction between two mercapto groups yields a cyclic disulfide and will be treated together with other problems of cyclization. Formation of intermolecular disulfide bridges connecting two identical chains requires merely dehydrogenation.

$$2 A\overset{\textstyle B}{\underset{SH}{\rule[0.5ex]{1em}{0.4pt}}} \longrightarrow \begin{array}{c} A\!-\!\!\!\top\!\!\!-B \\ S \\ | \\ S \\ A\!-\!\!\!\perp\!\!\!-B \end{array}$$

Various oxidizing agents such as potassium ferricyanide, iodine, 1,2-diiodoethane etc. have been proposed for this purpose but in most cases oxidation with air seems to be sufficient.

The bridging of two-non-identical chains is clearly a more complex task. Here, in addition to the desired mixed disulfide, oxidation yields the two symmetrical disulfides as well:

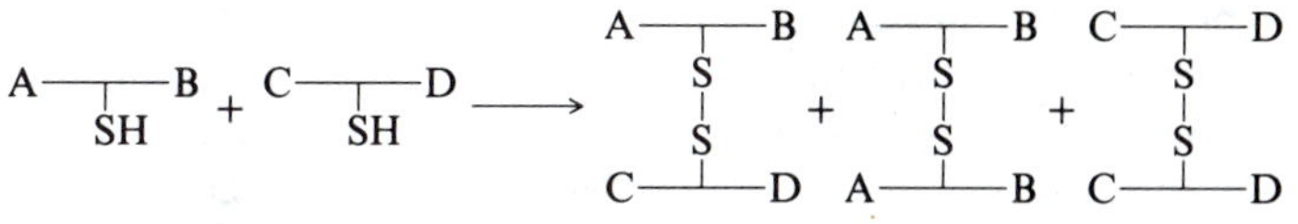

An analogous problem exists in the synthesis of peptides in which two identical chains are connected by two disulfide bridges. Oxidation of di-mercaptans affords a parallel and an antiparallel dimer and also oligomers or polymers:

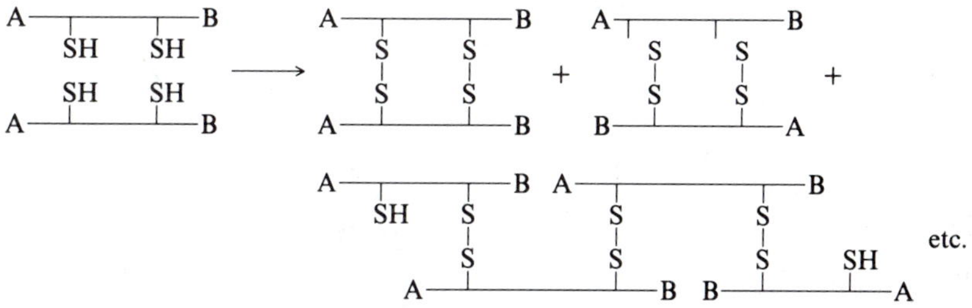

A directed formation of disulfides is also feasible. For instance the methoxycarbonylsulfenyl group (Brois et al. 1970) can be displaced from the sulfur atom of a cysteine containing peptide by the sulfhydryl group of the cysteine residue in a second peptide (Kamber 1973):

$$R-S-S-\overset{\overset{O}{\|}}{C}-OCH_3 + R'-SH \longrightarrow R-S-S-R' + COS + CH_3OH$$

This special sulfhydryl protecting group can be introduced via the reaction of an S-trityl derivative with methoxycarbonylsulfenyl chloride

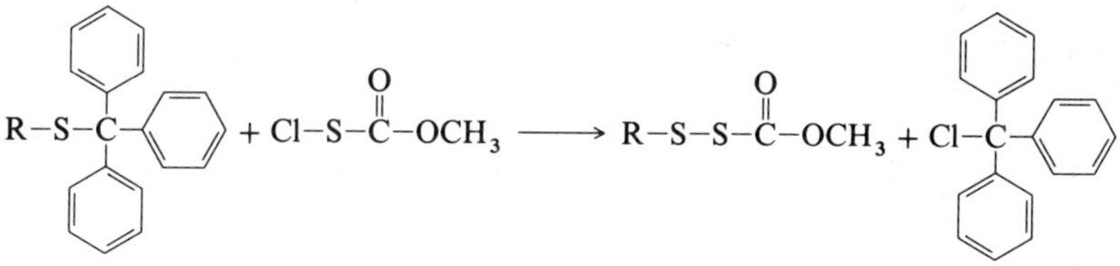

under mild conditions.

## 3. Cyclic Peptides

Peptides can form two kinds of cyclic structures. In *homodetic* cyclic peptides only peptide bonds participate in ring formation, while in *heterodetic* cyclic peptides a disulfide bridge, an ester (lactone) linkage or a thioether can play the role of the connecting piece between two points of the chain. In the synthesis of both classes of cyclic peptides a general problem is encountered. Cyclization requires two reactive groups within the same peptide chain and these can react with each other intramolecularly to produce the ring compound, but they might react in an intermolecular fashion as well to yield dimers or polymers:

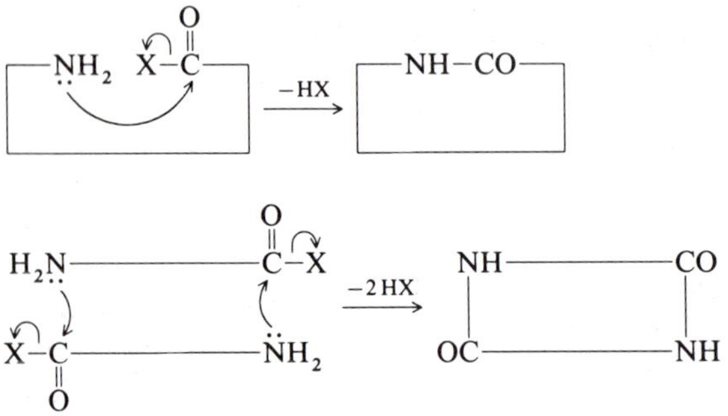

Production of larger rings can be more than a mere side reaction. When some interaction such as multiple hydrogen bonds holds the two activated chains together in an antiparallel combination

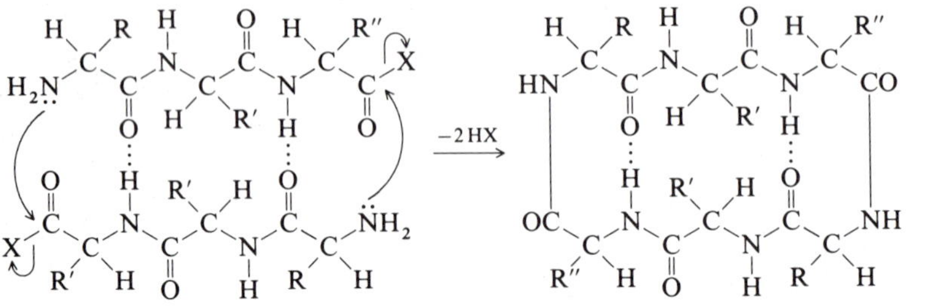

then the main product of the reaction will be the dimer rather than the monomer. Such cyclodimerization was observed (Schwyzer and Sieber 1958) in the synthesis of the cyclic antibiotic peptide gramicidin S:

$$2\ H-Val-\overset{\overset{\textstyle Tos}{|}}{Orn}-Leu-D-Phe-Pro-O-\langle\!\!\!\bigcirc\!\!\!\rangle-NO_2 \longrightarrow$$

$$\left[\begin{array}{c} \rightarrow Val-\overset{\overset{\textstyle Tos}{|}}{Orn}-Leu-D-Phe-Pro \\ \\ Pro-D-Phe-Leu-\underset{\underset{\textstyle Tos}{|}}{Orn}-Val \leftarrow \end{array}\right] + 2\,HO-\langle\!\!\!\bigcirc\!\!\!\rangle-NO_2$$

$$\overset{\overset{\textstyle Tos}{|}}{-Orn-} = CH_3-\langle\!\!\!\bigcirc\!\!\!\rangle-SO_2-\underset{\underset{\underset{\underset{-NH-CH-CO-}{\textstyle |}}{\textstyle CH_2}}{\overset{\textstyle |}{\underset{\textstyle CH_2}{|}}}}{NH}$$

Dimerization and polymerization can be prevented by the application of the *dilution principle* (Ziegler et al. 1933). In dilute solutions the rate of the bimolecular reaction is efficiently reduced while that of the simple cyclization, a unimolecular process, is unaffected. As a rule of thumb, cyclizations should be carried out at concentrations not exceeding $10^{-3}$ molar. This rule can be transgressed in some individual cases where a preferred quasi-cyclic conformation favors ring closure and impedes the bimolecular reaction.

In the synthesis of homodetic cyclic peptides ring closure requires the formation of a peptide bond between the amino group of the N-terminal and the carboxyl group of the C-terminal residue. Since no protection against racemization can be provided for the C-terminal residue, the latter must carefully be selected at the planning stage. If glycine is one of the amino acid constituents of the target compound, this is, of course the ideal choice. Proline, because its chiral integrity remains unaffected under most of the usual conditions of peptide synthesis, is next best. Proline was indeed chosen as the C-terminal residue of the open chain precursor in the already mentioned synthesis of gramicidine S. If neither glycine, nor proline occur in the sequence, then amino acids with no pronounced tendency for racemization, such as leucine or alanine can serve in this role. However, chiral purity is not the sole concern at this point. Residues prone to intramolecular reactions should not be selected, because the competing side reactions, such as glutarimide formation in activated glutamine residues adversely affect the yield of the desired cyclic peptide. Similarly, hindered amino acids, valine, isoleucine are not really suitable in the C-terminal position, because the reduced rate of cyclization will enhance the extent of side reactions, for instance O → N acyl migration in the O-acylisourea intermediate of carbodiimide mediated cyclizations:

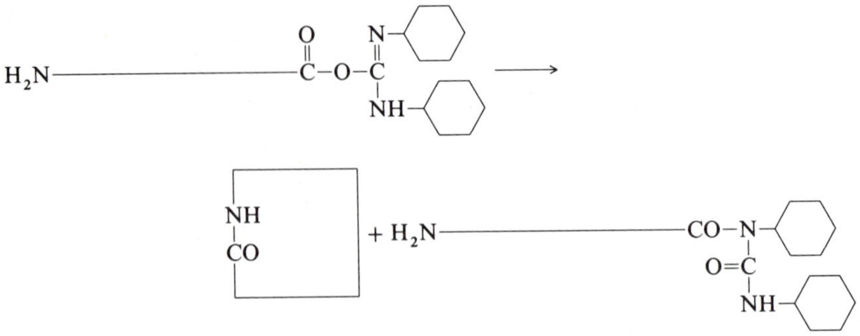

Similar difficulties should be expected in cyclization by the acid azide method, where Curtius rearrangement generates a cyclic urea-derivative instead of the desired cyclic peptide:

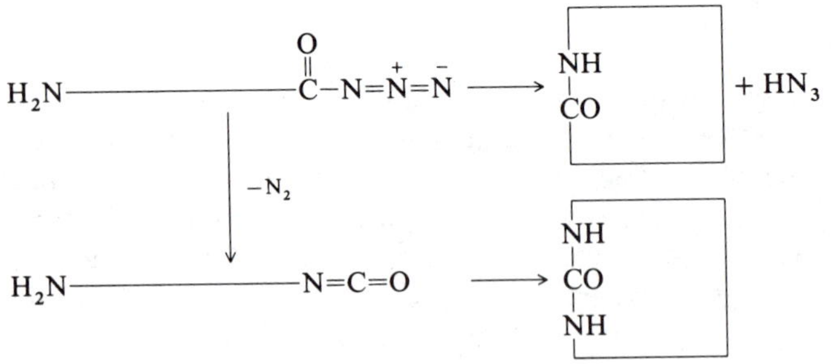

Nevertheless, satisfactory results were often obtained in cyclizations through the azide. The method rests on the large difference in the rate of conversion of the hydrazide to the azide and that of deamination of the (protonated) N-terminal residue by nitrous acid. Hence, with the calculated amount of sodium nitrite (or alkyl nitrite) rapid conversion of the hydrazide to the azide can take place without significant deamination at the N-terminus. Cyclization occurs on addition of base that deprotonates the amino group and thus permits its acylation. This addition, however, should be preceded by dilution to decrease the extent of intermolecular reactions

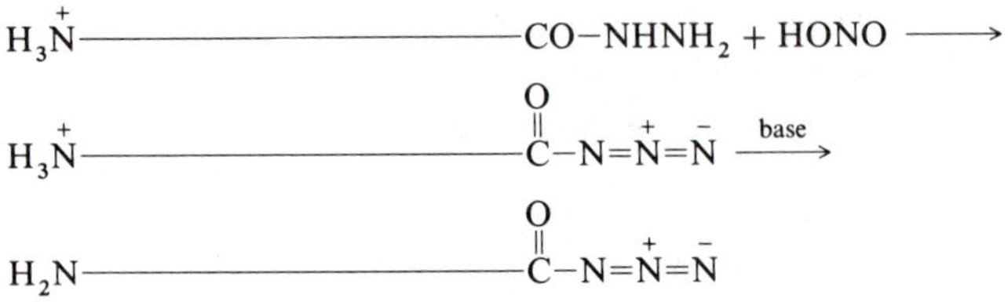

It can be said in general that separation of the step of activation from that of cyclization provides better control and often also better yields. A relatively simple realization of this principle is conversion of the partially blocked open chain precursor to an active ester followed by removal of the blocking group from the N-terminus:

Boc$-$NH$\longrightarrow$COOH + HO$-\langle\bigcirc\rangle-$NO$_2$ + DCC $\longrightarrow$

Boc$-$NH$\longrightarrow\overset{O}{\overset{\|}{C}}-$O$-\langle\bigcirc\rangle-$NO$_2$ $\xrightarrow{H^+}$

$\overset{+}{H_3N}\longrightarrow\overset{O}{\overset{\|}{C}}-$O$-\langle\bigcirc\rangle-$NO$_2$ $\xrightarrow[\text{2. base}]{\text{1. dilution}}$

H$_2$N$\longrightarrow\overset{O}{\overset{\|}{C}}-$O$-\langle\bigcirc\rangle-$NO$_2$ $\longrightarrow$

$\boxed{\begin{array}{c} \text{NH} \\ | \\ \text{CO} \end{array}}$ + HO$-\langle\bigcirc\rangle-$NO$_2$

In the preparation of *heterodetic* cyclic peptides ring closure might still take place at an amide bond. Thus a chain, which in addition to amides contains also an ester linkage (depsipeptides) can be cyclized by the methods described for homodetic peptides. It is equally possible, however, to form the ester bond in the process of cyclization. For the synthesis of peptide lactones the reagents used for amide bond formation can be applied, but less efficiently. The hydroxyl group is less nucleophilic than the amino group, hence ring closure is often accompanied by side reactions such as N-acylurea formation in the case of carbodiimides. This side reaction is suppressed by the addition of an auxiliary nucleophile such as 1-hydroxybenzotriazole and of course is absent when instead of carbodiimide a more unequivocal reagent for instance carbonyldiimidazole is used for activation. Solvents which enhance the nucleophilic character of the hydroxyl group can have a beneficial effect: pyridine might be the right choice. Base catalyzed transesterification of active esters is an additional possibility for lactone bond formation. Yet, all these approaches can lead to racemization at the activated residue since no adjoining amine-blocking group is present to prevent the loss of chiral purity. The imidazole generated in cyclization via carbonyldiimidazole or imidazole added as catalyst in the transesterification of active esters

$\boxed{}\overset{O}{\overset{\|}{C}}-$O$-\langle\bigcirc\rangle X$  HO$-$  $\underset{HN\diagdown\diagup N}{}$ $\longrightarrow$  $-$CO$-$O$-\boxed{}$ + HO$-\langle\bigcirc\rangle X$

is certainly undesirable in this respect. The more efficient catalyst 4-dimethyl-aminopyridine is also more harmful. Therefore lactone formation deserves further studies. For instance direct esterification with appropriate shifting of the ester equilibrium through azeotropic removal of water

$$\boxed{-COOH \quad HO-} \quad \xrightarrow[-H_2O]{\;\;\rightleftharpoons\;\;} \quad \boxed{-\overset{\overset{\displaystyle O}{\|}}{C}-O-}$$

might turn out to be a suitable method for the synthesis of peptide lactones.

Heterodetic cyclic peptides in which a disulfide bridge takes part in ring formation are obatined by the methods already discussed in connection with the construction of disulfide bridges between separate chains. Of course the dilution principle must be followed in order to diminish the extent of bimolecular reactions.

Several bicyclic compounds were found among microbial peptides, the toxic pentapeptide malformin being a relatively simple example:

$$
\begin{array}{c}
\text{D-Leu} \\
\diagup \qquad \diagdown \\
\text{Val} \qquad \qquad \text{Ile} \\
\diagdown \qquad \qquad \diagup \\
\text{D-Cys-D-Cys} \\
| \qquad\qquad | \\
\text{S}\text{------}\text{S}
\end{array}
$$

Synthesis of such materials can follow two alternative routes. Either the homodetic ring is secured first and the disulfide bridge is closed in the concluding step as in the synthesis of malformin (Bodanszky and Stahl 1974)

Ile−Cys(Bzl)−Cys(Bzl)−Val−Leu−N₃ $\longrightarrow$

$\rightarrow$ ⌐Ile−Cys(Bzl)−Cys(Bzl)−Val−Leu⌐ $\xrightarrow[\text{2. oxidation}]{\text{1. Na/NH}_3}$ $\rightarrow$⌐Ile−Cys−Cys−Val−Leu⌐

or, like in the synthesis of cyclo-cystine (Kamer 1971) first the heterodetic ring is formed by (partial) deblocking and oxidation to the disulfide:

$$
\begin{array}{l}
\text{Boc−NH−CH−CO−NH−CH−CO−OCH}_3 \\
\qquad\qquad | \qquad\qquad\qquad | \\
\qquad\qquad \text{CH}_2 \qquad\qquad\quad \text{CH}_2 \\
\text{CH}_3\text{CO−NH−CH}_2\text{−S} \qquad \text{S−CH}_2\text{−NH−COCH}_3
\end{array}
\qquad \xrightarrow{\text{I}_2/\text{CH}_3\text{OH}}
$$

$$
\begin{array}{l}
\text{Boc−NH−CH−CO−NH−CH−CO−OCH}_3 \\
\qquad\qquad | \qquad\qquad\qquad | \\
\qquad\qquad \text{CH}_2\text{—S—S—CH}_2
\end{array}
\qquad \xrightarrow[\text{2. heating}]{\text{1. HCOOH}}
\qquad
\begin{array}{c}
\text{CH} \\
\text{HN}\diagup\;\diagdown\text{CO} \\
| \quad \text{S} \quad | \\
| \quad \text{``S} \quad | \\
\text{OC}\diagdown\;\diagup\text{NH} \\
\text{CH}
\end{array}
$$

Similarly a thioether was first prepared in the synthesis of the toxic mushroom peptide phalloidin and the homodetic ring was closed only subsequently

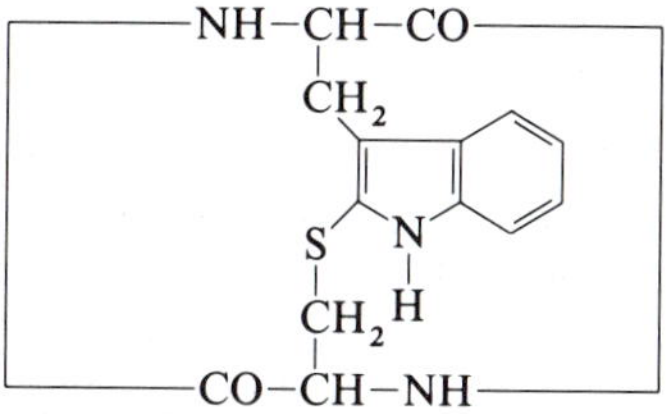

The presence of a D-amino acid in the sequence of these peptides or the alternation of L and D residues (cf. malformin) results in preferred conformations (reverse turns) that facilitate ring closure. Geometries conducive to cyclization are generated also by the presence of a proline residue in the sequence, while glycine, by the absence of an impeding side chain can be similarly helpful. The cyclization promoting effect of bulky side chains, particularly valine and isoleucine with branching at the $\beta$-carbon atom, can not be excluded.

## 4. Peptides Containing Non-amino Acid Constituents

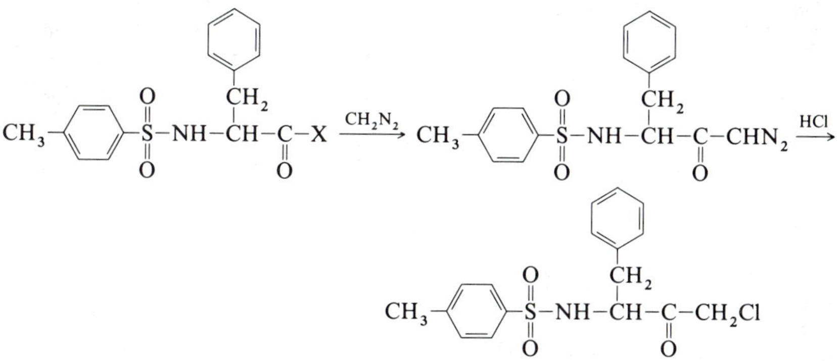

Short peptide chains in which one of the amino acid residues is replaced by a moiety related (or sometimes unrelated) to it appear with increased frequency in the literature. In the center of interest are enzyme inhibitors. Synthesis of these compounds is obviously determined by the reactivity of the unusual constituent and therefore it is hardly possible to set rules for such endeavours. We must confine the discussions therefore to generalities. For example chloromethylketones derived from certain amino acids are irreversible inhibitors which not merely interact but indeed react with the active site of an enzyme with the formation of a covalent bond. Chloromethyl ketones are synthesized by activation of the blocked amino acid followed by reaction with diazomethane to yield a diazoketone. This is transformed to the corresponding chloromethyl ketone by HCl in an organic solvent. From p-toluenesulfonyl-L-phenylalanine a chloromethyl ketone is obtained which selectively and irreversibly blocks chymotrypsin. If, in order to increase specificity, addition of a short peptide with distinct affinity to the active site of a chymotrypsin-like enzyme is needed, then the toluenesulfonyl (tosyl) protecting group is unsuited because its removal prior to chain lengthening toward the N-terminus requires reduction with sodium in liquid ammonia and under the conditions of the deblocking procedure the chloromethyl group is destroyed. Replacement of the tosyl group by the 9-fluorenyloxycarbonyl (Fmoc) group would create a similar obstacle: deblocking with a secondary amine is obviously not feasible without alkylation of the reagent by the reactive chloromethylketone group. Hence protecting groups should be selected that can be removed by acidolysis, for instance the tert.butyloxycarbonyl (Boc) group. The benzyloxycarbonyl (Z) group is again not fully acceptable because during hydrogenolysis the chlorine atom is replaced by hydrogen and acidolysis with HBr in acetic acid gives low yields. Prolonged treatment with trifluoroacetic acid is somewhat better but not entirely satisfactory. Analogous problems arise in the synthesis of peptides which contain a nitrogen mustard grouping; this is also relatively insensitive to acids but reacts with bases. A careful scrutiny of the reactivity of the inhibitor moiety is necessary before a scheme for synthesis can be developed. In the case of serious incompatibilities alternative routes such as assembling a peptide with phenylalanine at its C-terminus and converting it to the chloromethyl ketone at the conclusion of the synthesis may have to be considered.

## 5. Semisynthesis (Partial Synthesis)

The intriguing possibility to utilize parts of a protein in the synthesis of larger peptides and even proteins stimulated considerable research. Selective fragmentation at well defined peptide bonds is possible with proteolytic enzymes and also by some chemical methods such as cleavage of methionine containing chains with cyanogen bromide. Since it is often problematic to provide the fragments with appropriate blocking groups, masking of at least the amine function is preferably carried out prior to fragmentation. Special blocking methods are available for this purpose: trifluoroacetylation, acetimidylation and maleylation, all removable under very mild conditions:

$$CF_3\overset{\overset{\text{O}}{\|}}{C}-S-CH_2CH_3 + \quad \underset{\underset{-NH-\overset{|}{C}H-CO-}{(\overset{|}{C}H_2)_4}}{\overset{H_2N}{|}} \longrightarrow \quad \underset{\underset{-NH-\overset{|}{C}H-CO-}{(\overset{|}{C}H_2)_4}}{\overset{CF_3CO-NH}{|}} + C_2H_5SH$$

$$CH_3-\overset{\overset{+}{\overset{NH_2}{\|}}}{C}-OCH_3 + \quad \underset{\underset{-NH-\overset{|}{C}H-CO-}{(\overset{|}{C}H_2)_4}}{\overset{H_2N}{|}} \longrightarrow \quad \underset{\underset{-NH-\overset{|}{C}H-CO-}{(\overset{|}{C}H_2)_4}}{\overset{\overset{+}{\overset{NH_2}{\|}}}{CH_3-C-NH}} + CH_3OH$$

$$
\underset{\substack{\text{O} \\ \text{(CH}_2)_4 \\ -\text{NH}-\text{CH}-\text{CO}-}}{\overset{\text{O}}{\underset{\text{O}}{\overset{\overset{\displaystyle \text{C}}{\|}}{\underset{\underset{\text{C}}{\|}}{\bigg[}}}\text{O}}} + \quad \underset{\substack{\text{(CH}_2)_4 \\ -\text{NH}-\text{CH}-\text{CO}-}}{\text{H}_2\text{N}} \longrightarrow \underset{\substack{\text{(CH}_2)_4 \\ -\text{NH}-\text{CH}-\text{CO}-}}{\overset{\text{CH}-\text{COOH}}{\underset{\|}{\text{CH}-\text{CO}-\text{NH}}}}
$$

Activation of the newly formed carboxyl group can be simplified if enzymatic cleavage, e.g. with trypsin, is carried out in the presence of excess hydrazine: the peptide hydrazide is obtained

$$
\begin{array}{c}
\text{NH} \\
\| \\
\text{NH}-\text{C}-\text{NH}_2 \\
| \\
\text{CH}_2 \\
| \\
\text{CH}_2 \\
| \\
\text{CH}_2 \\
| \\
-\text{NH}-\text{CH}-\text{CO}-\text{NH}-\text{CHR}-\text{CO}-
\end{array}
\xrightarrow[\text{H}_2\text{NNH}_2]{\text{trypsin}}
\begin{array}{c}
\text{NH} \\
\| \\
\text{NH}-\text{C}-\text{NH}_2 \\
| \\
\text{CH}_2 \\
| \\
\text{CH}_2 \\
| \\
\text{CH}_2 \\
| \\
-\text{NH}-\text{CH}-\text{CO}-\text{NHNH}_2
\end{array}
\quad + \; \text{H}_2\text{N}-\text{CHR}-\text{CO}-
$$

and is then activated in the form of the azide by treatment with nitrous acid or alkyl nitrites.

The complex methodology of semisynthesis was treated in detail by Offord (1980).

# References

Bodanszky, M.: Ann. New York Acad. Sci. *88*, 655 (1960)

Bodanszky, M., du Vigneaud, V.: J. Amer. Chem. Soc. *81*, 5688 (1959)

Bodanszky, M., Ondetti, M. A., Levine, S. D., Williams, N. J.: J. Amer. Chem. Soc. *89*, 6753 (1967)

Bodanszky, M., Stahl, G. L.: Proc. Nat. Acad. Sci. USA *71*, 2791 (1974)

Brois, S. J., Pilot, J. F., Barnum, H. W.: J. Amer. Chem. Soc. *92*, 2791 (1970)

du Vigneaud, V., Ressler, C., Swan, J. M., Roberts, C. W., Katsoyannis, P. G., Gordon, S.: J. Amer. Chem. Soc. *75*, 4879 (1953)

Kamber, B.: Helv. Chim. Acta *54*, 927 (1971)

Kamber, B.: Helv. Chim. Acta *56*, 1370 (1973)

Kemp, D. S., Bernstein, Z. W., McNeil, G. N.: J. Org. Chem. *39*, 2831 (1974)

Miyoshi, M., Kimura, T., Sakakibara, S.: Bull. Chem. Soc. Jpn. *43*, 2941 (1970)

Offord, R. E.: Semisynthetic Proteins, John Wiley, New York, 1980

Sakakibara, S., Kishida, Y., Kikuchi, Y., Sakai, R., Kakiuchi, K.: Bull. Chem. Soc. Jpn. *41*, 1273 (1968)

Schwyzer, R., Sieber, P.: Helv. Chim. Acta *41*, 2186 (1958)

Wieland, Th.: Peptides of Poisonous Amanita Mushrooms, Springer Verlag, New York, 1986

Yajima, H., Fujii, N.: J. Amer. Chem. Soc. *103*, 5867 (1981)

Ziegler, K., Eberle, H., Ohlinger, H.: Liebigs Ann. Chem. *504*, 95 (1933)

## Additional Sources

Jones, J. H.: Sequential Peptide Synthesis in Chemistry and Biochemistry of Amino
    Acids, Peptides and Proteins, Vol. *4*, Weinstein, B. ed., pp. 29–63, Dekker, New York
    1977

Kricheldorf, H. R.: α-Aminoacid-N-Carboxy-Anhydrides and Related Heterocycles,
    Springer Verlag, Berlin-Heidelberg, 1987

Roberts, D. C., Velaccio, F.: Unusual Amino Acids in Peptide Synthesis in The Peptides,
    Vol. 5, Gross, E., Meienhofer, J. eds., pp. 341–449, Academic Press, New York 1983

Sheppard, R. C.: Partial Synthesis of Peptides and Proteins in The Peptides Vol. 2, Gross,
    E., Meienhofer, J. eds., pp. 441–484, Academic Press, New York 1980

# X. Solid Phase Peptide Synthesis

By the conventional methods of organic synthesis the preparation of peptides containing more than just a few amino acids is an arduous task. Introduction of blocking groups, coupling reactions and deprotection steps entail a large number of operations such as washing the reaction mixtures neutral after coupling, precipitation or crystallization of intermediates, collecting solid products by filtration or centrifugation followed by drying etc. Thus, synthesis of peptide chains containing dozens of residues requires an almost heroic effort and proteins, even small ones, can be made by tour de force but certainly not routinely. The need for facilitation of the process was obvious for some time. The stepwise strategy, demonstrated in a novel synthesis of oxytocin (Bodanszky and du Vigneaud 1959) was, because of the repetitiveness of the operations conducive to experimentation with techniques suitable for mechanization and automation of chain building. Attachment of the (N-blocked) C-terminal residue to an insoluble polymeric support (a "resin") followed by deprotection and acylation of the exposed amino group with the penultimate residue and continuation of the procedure by similar cycles of deprotection and incorporation absolve the practitioner from handling filters and separatory funnels, from washing and drying intermediates etc. Excess starting material and reagents as well as byproducts of the reactions are eliminated simply by washing the peptidyl polymer with appropriately selected solvents. The peptide is separated from the support only after completion of chain building. This new concept was realized by R. B. Merrifield (1963) in the synthesis of a tetrapeptide L-leucyl-L-alanyl-glycyl-L-valine. The insoluble support developed for this purpose, the Merrifield resin, is still one of the mainstays of solid phase peptide synthesis. It was obtained through the chloromethylation of a styrene-divinylbenzene copolymer

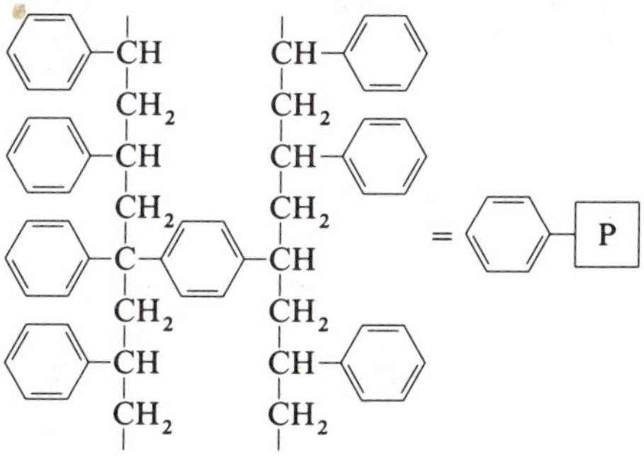

$$\langle\!\!\bigcirc\!\!\rangle\!-\!\boxed{P} + ClCH_2OCH_3 \xrightarrow{ZnCl_2} ClCH_2\!-\!\langle\!\!\bigcirc\!\!\rangle\!-\!\boxed{P} + CH_3OH$$

Anchoring of the first amino acid (the C-terminal residue of the chain to be built) was achieved through reaction of the chloromethylated resin with the triethylammonium salt of the benzyloxycarbonyl amino acid

$$\langle\!\!\bigcirc\!\!\rangle\!-\!CH_2O\!-\!CO\!-\!NH\!-\!CHR\!-\!COO^- + ClCH_2\!-\!\langle\!\!\bigcirc\!\!\rangle\!-\!\boxed{P} \longrightarrow$$

$$\langle\!\!\bigcirc\!\!\rangle\!-\!CH_2O\!-\!CO\!-\!NH\!-\!CHR\!-\!CO\!-\!OCH_2\!-\!\langle\!\!\bigcirc\!\!\rangle\!-\!\boxed{P} + Cl^-$$

Removal of the benzyloxycarbonyl group with HBr in acetic acid yielded the aminoacyl resin which could be acylated with the next amino acid to be incorporated, applied, of course, in N-blocked and carboxyl-activated form

$$\langle\!\!\bigcirc\!\!\rangle\!-\!CH_2O\!-\!CO\!-\!NH\!-\!CHR\!-\!CO\!-\!OCH_2\!-\!\langle\!\!\bigcirc\!\!\rangle\!-\!\boxed{P} \xrightarrow[\text{2. N(C}_2\text{H}_5)_3]{\text{1. HBr/AcOH}}$$

$$H_2N\!-\!CHR\!-\!CO\!-\!OCH_2\!-\!\langle\!\!\bigcirc\!\!\rangle\!-\!\boxed{P} \xrightarrow{Z-NH-CHR'-CO-X}$$

$$\langle\!\!\bigcirc\!\!\rangle\!-\!CH_2O\!-\!CO\!-\!NH\!-\!CHR'\!-\!CO\!-\!NH\!-\!CHR\!-\!CO\!-\!OCH_2\!-\!\langle\!\!\bigcirc\!\!\rangle\!-\!\boxed{P}$$

While it was possible to cleave the benzyloxycarbonyl group without affecting in a major way the benzyl ester-like grouping linking the C-terminal residue to the polymer, this selectivity was not sufficient for practical purposes: in each deprotection step a small but not negligible part of the amino acid or peptide was cleaved from the polymeric support and lost. To improve selectivity the aromatic nuclei in the polymer were nitrated: p-nitrobenzyl esters show considerable resistance toward acids. This modification however, while improving selectivity in deprotection, made it necessary to use saponification with alkali, a somewhat ambiguous method, for cleavage of the completed peptide from the resin

$$-NH\!-\!CHR''\!-\!CO\!-\!NH\!-\!CHR'\!-\!CO\!-\!NH\!-\!CHR\!-\!CO\!-\!OCH_2\!-\!\underset{O_2N}{\langle\!\!\bigcirc\!\!\rangle}\!-\!\boxed{P} \xrightarrow{OH^-}$$

$$-NH\!-\!CHR''\!-\!CO\!-\!NH\!-\!CHR'\!-\!CO\!-\!NH\!-\!CHR\!-\!COO^- + HOCH_2\!-\!\underset{O_2N}{\langle\!\!\bigcirc\!\!\rangle}\!-\!\boxed{P}$$

Therefore further improvement was sought and found in the application of tert.butyloxycarbonyl rather than benzyloxycarbonyl amino acids. In the revised procedure (Merrifield 1964) nitration of the resin became superfluous, because the relatively mild acidic conditions needed for deprotection after each coupling step, for instance N HCl in acetic acid or trifluoroacetic acid diluted with dichloromethane do not significantly cleave the anchoring bond. In the initial studies of Merrifield dicyclohexylcarbodiimide was used as coupling reagent and it remains the activating reagent of choice in most laboratories. The solid phase approach is illustrated in Scheme 3.

**Scheme 3**

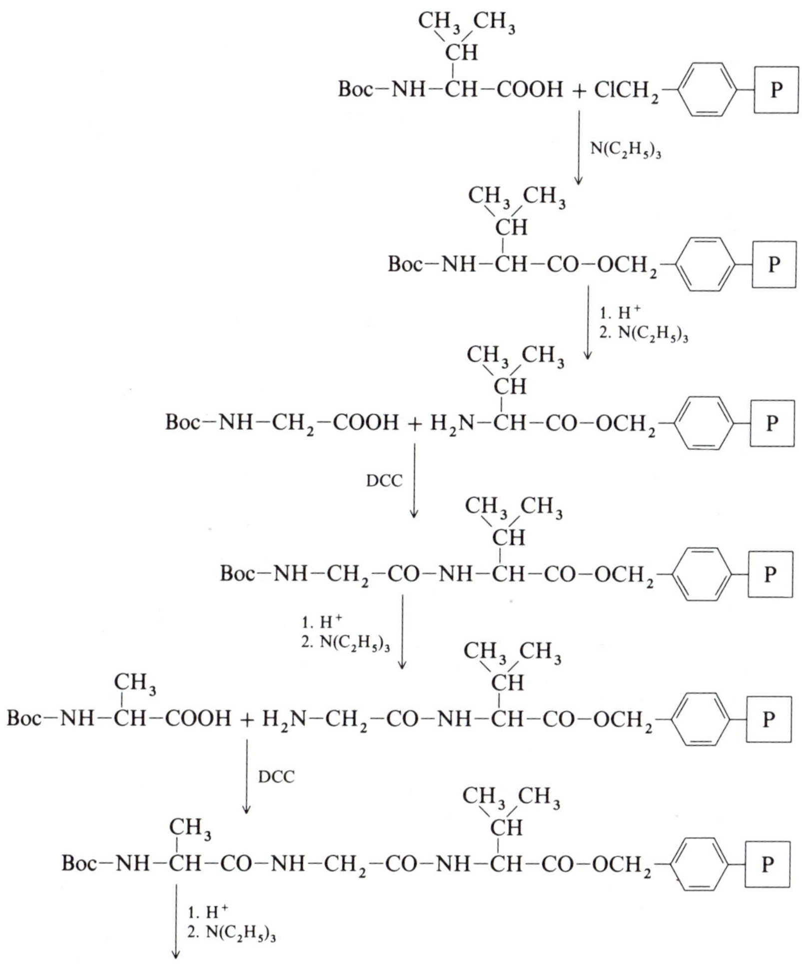

**Scheme 3** (continued)

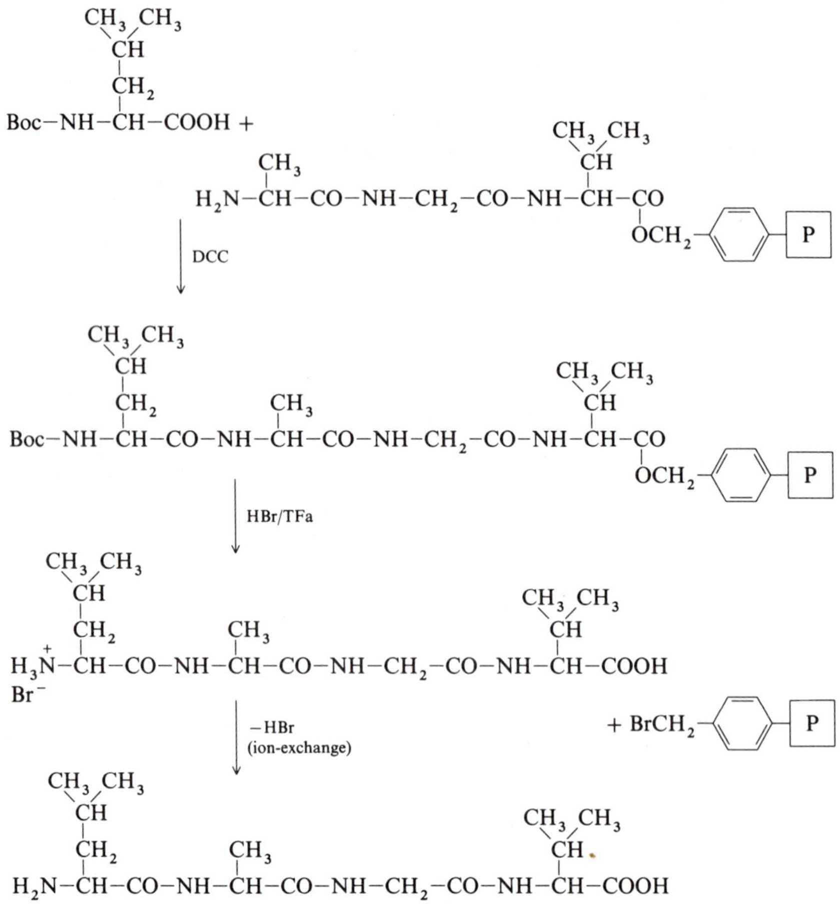

$$\begin{array}{c} CH_3 \diagdown \diagup CH_3 \\ CH \\ | \\ CH_2 \\ | \\ Boc-NH-CH-COOH \end{array} +$$

$$\begin{array}{c} CH_3 \\ | \\ H_2N-CH-CO-NH-CH_2-CO-NH-CH-CO \end{array} \quad \begin{array}{c} CH_3 \diagdown \diagup CH_3 \\ CH \\ | \\ \\ OCH_2-\!\!\boxed{\phantom{aa}}\!\!-\boxed{P} \end{array}$$

| DCC

$$\begin{array}{c} CH_3 \diagdown \diagup CH_3 \\ CH \\ | \\ CH_2 \qquad\qquad CH_3 \qquad\qquad\qquad CH \\ | \qquad\qquad\qquad | \qquad\qquad\qquad\qquad\qquad | \\ Boc-NH-CH-CO-NH-CH-CO-NH-CH_2-CO-NH-CH-CO \\ \qquad\qquad\qquad\qquad\qquad\qquad\qquad\qquad\qquad OCH_2-\!\!\boxed{\phantom{aa}}\!\!-\boxed{P} \end{array}$$

| HBr/TFa

$$\begin{array}{c} CH_3 \diagdown \diagup CH_3 \\ CH \\ | \\ CH_2 \qquad\qquad CH_3 \qquad\qquad\qquad CH \\ | \qquad\qquad\qquad | \qquad\qquad\qquad\qquad\qquad | \\ \overset{+}{H_3}N-CH-CO-NH-CH-CO-NH-CH_2-CO-NH-CH-COOH \\ Br^- \end{array}$$

| −HBr     + BrCH$_2$−$\boxed{\phantom{aa}}$−$\boxed{P}$
| (ion-exchange)

$$\begin{array}{c} CH_3 \diagdown \diagup CH_3 \\ CH \\ | \\ CH_2 \qquad\qquad CH_3 \qquad\qquad\qquad CH . \\ | \qquad\qquad\qquad | \qquad\qquad\qquad\qquad\qquad | \\ H_2N-CH-CO-NH-CH-CO-NH-CH_2-CO-NH-CH-COOH \end{array}$$

Leu−Ala−Gly−Val

It is quite possible to use the versatile solid phase method in segment condensation as well, sometimes with considerable advantage. For instance isolation of difficult-to-handle intermediates can be circumvented by preparing them through a resin-bound amine-component. More importantly, if this component is present as the salt of the acid used in the preceding deprotection, the free amine can be secured by merely washing the peptidyl-resin with a solution of triethylamine in dichloromethane. Yet, the ultimate aim of the solid phase method, *mechanization and automation of chain building* can only be attained if the stepwise strategy is followed. Hence, in the following sections only this generally applied procedure, the stepwise incorporation of single amino acid residues will be discussed.

## A. The Insoluble Polymeric Support and the Bond Linking the Peptide to the Resin

The Merrifield resin, chloromethylated styrene-divinylbenzene copolymer, proved itself in numerous syntheses but was also amenable to further development. Reduction of the divinylbenzene content of the polymerization mixture from 2 to 1 % resulted in a resin that swells better in the solvents used in the process, such as dichloromethane or dimethylformamide. This improvement, bought at the price of decreased mechanical stability, allows faster penetration (by diffusion) of the reactants. Thus, higher concentration of the acylating agent around the amino groups can be reached at a given time. The latter, surrounded by the matrix of the polymer, are more readily available in a well swollen resin whereby the rate-limiting effect of diffusion is considerably reduced. Also, in the 1 % cross-linked polymer it is less likely that a small fraction of the molecules of the amino component remains hidden in certain areas of the matrix and is unavailable for acylation by activated derivatives of hindered amino acids such as valine or isoleucine. Of course, cleavage of blocking groups and conversion of salts to free amines are similarly facilitated.

In addition to physical improvements the chemistry of the chloromethylated resin could also be improved. Displacement of the chlorine atom by the acetate ion takes place on heating the resin with a solution of potassium acetate in a high boiling alcohol. Saponification of the ester with an alcoholic solution of potassium hydroxide or ammonolysis in methanol yields the hydroxymethyl-resin, a polymer-bound form of benzyl alcohol:

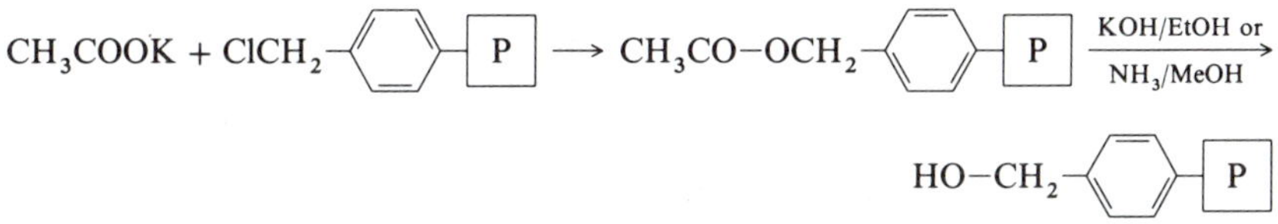

$$CH_3COOK + ClCH_2-\langle\rangle-\boxed{P} \longrightarrow CH_3CO-OCH_2-\langle\rangle-\boxed{P} \quad \xrightarrow[NH_3/MeOH]{KOH/EtOH\ or}$$

$$HO-CH_2-\langle\rangle-\boxed{P}$$

Anchoring of the first amino acid by esterification with the hydroxymethyl-resin can be accomplished with the help of condensing agents such as dicyclohexylcarbodiimide or preferably with carbonyldiimidazole. A more convenient method, the imidazole catalyzed transesterification of active esters

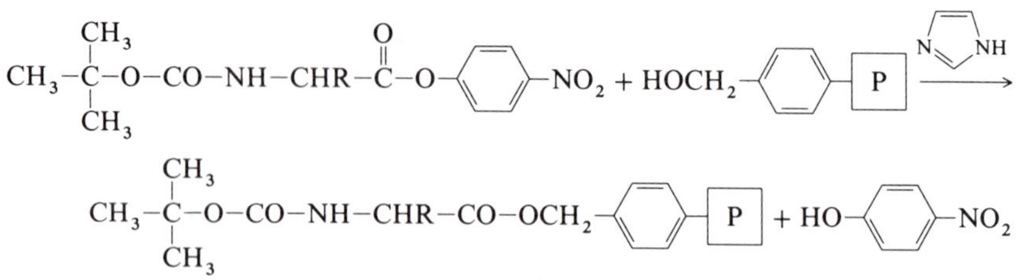

$$CH_3-\underset{\underset{CH_3}{|}}{\overset{\overset{CH_3}{|}}{C}}-O-CO-NH-CHR-\overset{\overset{O}{\|}}{C}-O-\langle\ \rangle-NO_2 + HOCH_2-\langle\ \rangle-\boxed{P} \longrightarrow$$

$$CH_3-\underset{\underset{CH_3}{|}}{\overset{\overset{CH_3}{|}}{C}}-O-CO-NH-CHR-CO-OCH_2-\langle\ \rangle-\boxed{P} + HO-\langle\ \rangle-NO_2$$

must be performed in non-polar solvents, preferably in toluene, to avoid racemization. The very efficient catalyst 4-dimethylaminopyridine will cause racemization even under such conditions. The advantages of the hydroxymethyl-resin [1] over the chloromethylated polymer are not negligible. In the anchoring step the chloromethyl group can alkylate nucleophilic side chain functions, for instance the imidazole nucleus in histidine or the thioether sulfur atom in methionine

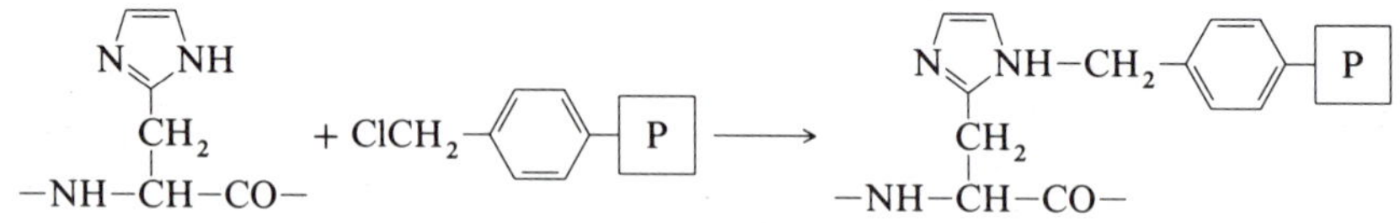

$$\underset{\underset{-NH-CH-CO-}{\underset{|}{CH_2}}}{N\diagdown NH} + ClCH_2-\langle\ \rangle-\boxed{P} \longrightarrow \underset{\underset{-NH-CH-CO-}{\underset{|}{CH_2}}}{N\diagdown NH-CH_2-\langle\ \rangle-\boxed{P}}$$

---

[1] It should be noted that a hydroxymethylated insoluble support was prepared through copolymerization of 4-hydroxymethylstyrene, styrene and divinylbenzene and also via the reduction of a carboxyl-resin with LiAlH$_4$ by Letsinger and his associates (1963), who simultaneously and independently from Merrifield initiated solid phase peptide synthesis. In their approach the polymer was treated with phosgene and the polymeric analog of benzyl chlorocarbonate thus obtained used for the acylation of an amino acid:

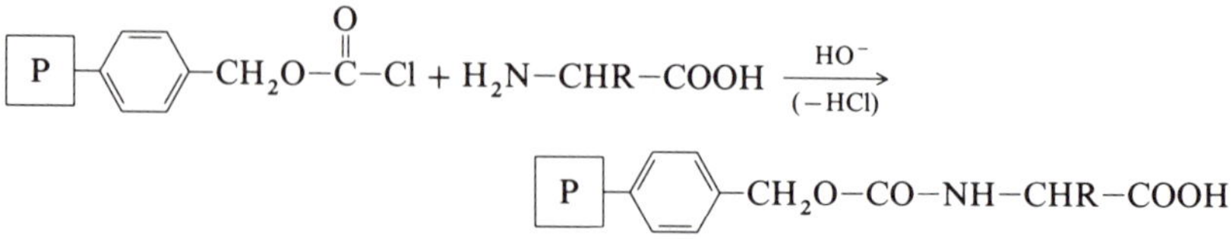

$$\boxed{P}-\langle\ \rangle-CH_2O-\overset{\overset{O}{\|}}{C}-Cl + H_2N-CHR-COOH \xrightarrow[(-HCl)]{HO^-}$$

$$\boxed{P}-\langle\ \rangle-CH_2O-CO-NH-CHR-COOH$$

The resin bound amino acid was designed to serve as the N-terminal of a peptide to be built through stepwise chain lengthening. The difficulties inherent in this strategy, noted in the preceding chapter, prevented general acceptance of the Letsinger version of solid phase peptide synthesis, but development of coupling methods free of racemization could lead to its revival.

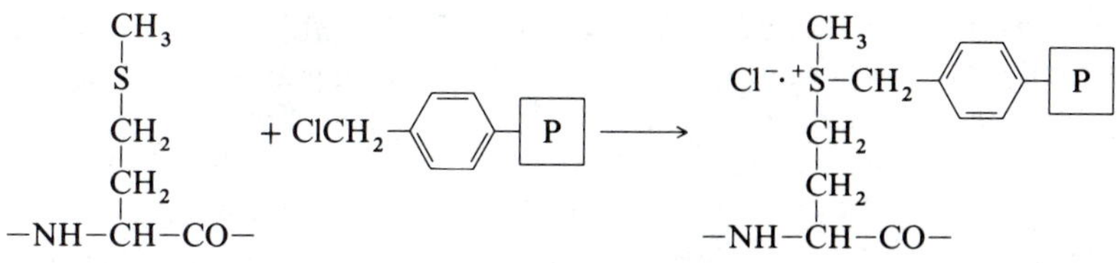

$$\begin{array}{c} CH_3 \\ | \\ S \\ | \\ CH_2 \\ | \\ CH_2 \\ | \\ -NH-CH-CO- \end{array} \quad + \; ClCH_2-\!\!\left\langle\!\!\bigcirc\!\!\right\rangle\!\!-\boxed{P} \quad \longrightarrow \quad \begin{array}{c} CH_3 \\ | \\ Cl^{-}\cdot{}^{+}S-CH_2-\!\!\left\langle\!\!\bigcirc\!\!\right\rangle\!\!-\boxed{P} \\ | \\ CH_2 \\ | \\ CH_2 \\ | \\ -NH-CH-CO- \end{array}$$

and also tertiary bases, such as triethylamine,

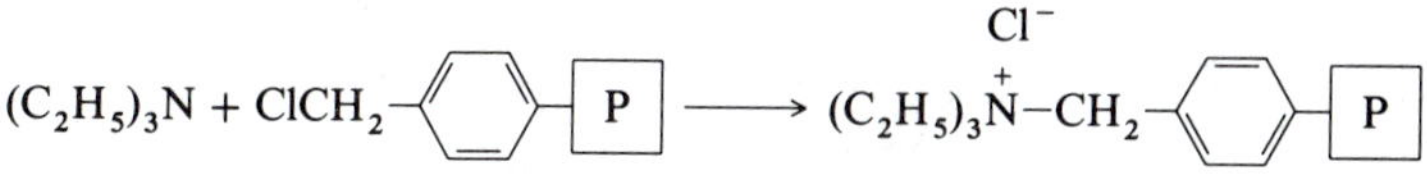

$$(C_2H_5)_3N + ClCH_2-\!\!\left\langle\!\!\bigcirc\!\!\right\rangle\!\!-\boxed{P} \quad \longrightarrow \quad (C_2H_5)_3\overset{+}{N}-CH_2-\!\!\left\langle\!\!\bigcirc\!\!\right\rangle\!\!-\boxed{P} \quad \overset{Cl^-}{}$$

used in the process.

A further development of the hydroxymethyl-resin concept can be recognized in polymeric supports such as

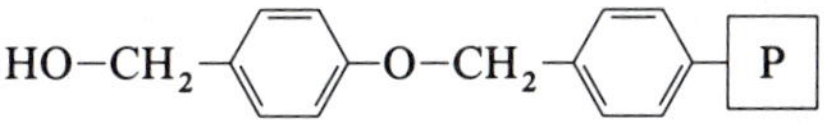

$$HO-CH_2-\!\!\left\langle\!\!\bigcirc\!\!\right\rangle\!\!-O-CH_2-\!\!\left\langle\!\!\bigcirc\!\!\right\rangle\!\!-\boxed{P}$$

Here the p-alkoxy substituent of the benzyl alcohol moiety enhances the acid sensitivity of the ester linkage between peptide and resin and thus the anchoring bond can be cleaved with trifluoroacetic acid. Consequently the blocking groups applied for the protection of the α-amino function must be much more acid-sensitive, like the biphenylylisopropyloxycarbonyl (Bpoc) group:

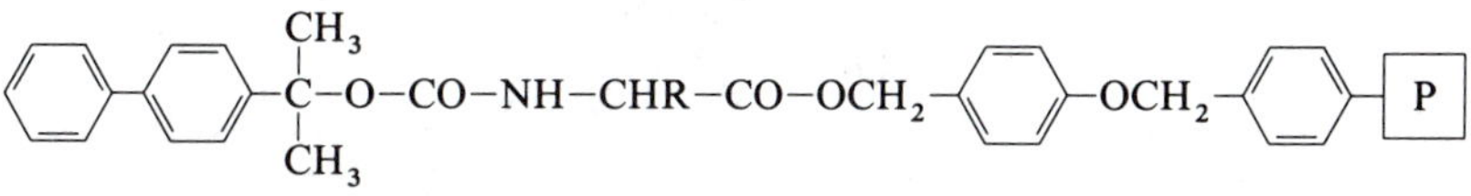

$$\left\langle\!\!\bigcirc\!\!\right\rangle\!\!-\!\!\left\langle\!\!\bigcirc\!\!\right\rangle\!\!-\!\!\begin{array}{c} CH_3 \\ | \\ C \\ | \\ CH_3 \end{array}\!\!-O-CO-NH-CHR-CO-OCH_2-\!\!\left\langle\!\!\bigcirc\!\!\right\rangle\!\!-OCH_2-\!\!\left\langle\!\!\bigcirc\!\!\right\rangle\!\!-\boxed{P}$$

An orthogonal and therefore more reliable combination is offered by the base sensitive 9-fluorenylmethyloxycarbonyl (Fmoc) group:

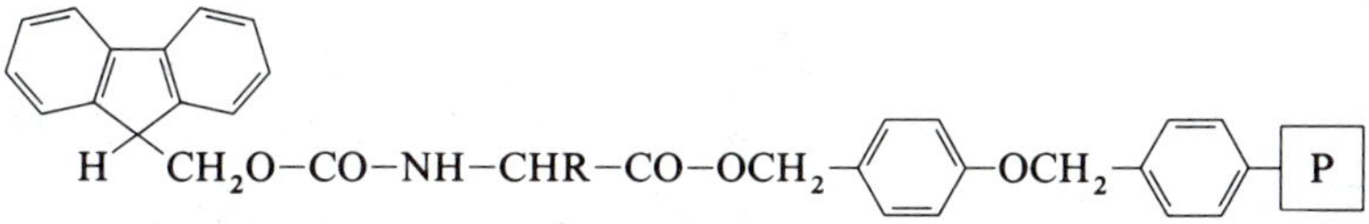

$$\underset{H}{\text{(fluorenyl)}}\!\!\begin{array}{c} \\ CH_2O-CO-NH-CHR-CO-OCH_2-\!\!\left\langle\!\!\bigcirc\!\!\right\rangle\!\!-OCH_2-\!\!\left\langle\!\!\bigcirc\!\!\right\rangle\!\!-\boxed{P} \end{array}$$

At this point we have to return to the problem of stability of the anchoring ester bond. Cleavage of the Boc blocking group with dilute HCl in acetic acid or with a mixture of trifluoroacetic acid and dichloromethane does not affect significantly the benzyl ester type linkage. Yet, even a minute loss of peptide from the resin is disturbing in the synthesis of long chains because acidolytic removal of $\alpha$-amine protecting groups is many times repeated. Therefore, as the target compounds increased in length, anchoring bonds with improved resistance to acids became once more desirable. From the numerous proposals advanced for this purpose the phenylacetamidomethyl derivatives (PAM resins)

$$R-CO-OCH_2-\langle\ \rangle-CH_2CO-NH-CH_2-\langle\ \rangle-\boxed{P}$$

gained considerable popularity. A practical method leading to the appropriate starting material is coupling a phenylacetic acid derivative comprising the C-terminal amino acid to a modified polymer, the aminomethyl resin

$$CH_3-\underset{\underset{CH_3}{|}}{\overset{\overset{CH_3}{|}}{C}}-O-CO-NH-CHR-CO-OCH_2-\langle\ \rangle-CH_2-COOH\ +$$

$$H_2N-CH_2-\langle\ \rangle-\boxed{P}\ \xrightarrow{\langle\ \rangle-N=C=N-\langle\ \rangle}$$

$$CH_3-\underset{\underset{CH_3}{|}}{\overset{\overset{CH_3}{|}}{C}}-O-CO-NH-CHR-CO-OCH_2-\langle\ \rangle-CH_2-CO-NH\underset{CH_2-\langle\ \rangle-\boxed{P}}{|}$$

The aminomethyl copolymer is readily obtained from the chloromethyl resin

$$\overset{CO}{\underset{CO}{\langle\ \rangle}}NK\ +\ ClCH_2-\langle\ \rangle-\boxed{P}\ \longrightarrow\ \overset{CO}{\underset{CO}{\langle\ \rangle}}N-CH_2-\langle\ \rangle-\boxed{P}$$

$$\xrightarrow{H_2NNH_2}\ H_2N-CH_2-\langle\ \rangle-\boxed{P}\ +\ \overset{CO}{\underset{CO}{\langle\ \rangle}}\overset{NH}{\underset{NH}{|}}$$

or directly from the commercially available styrene-divinylbenzene copolymer:

$$\overset{CO}{\underset{CO}{\langle\ \rangle}}N-CH_2OH\ +\ \langle\ \rangle-\boxed{P}\ \xrightarrow[(CF_3SO_3H)]{H^+}\ \overset{CO}{\underset{CO}{\langle\ \rangle}}N-CH_2-\langle\ \rangle-\boxed{P}$$

$$\xrightarrow{H_2NNH_2}\ H_2N-CH_2-\langle\ \rangle-\boxed{P}\ +\ \overset{CO}{\underset{CO}{\langle\ \rangle}}\overset{NH}{\underset{NH}{|}}$$

In such sophisticated functionalization of the polymeric support, in addition to the "handle" which allows the attachment of the C-terminal amino acid, also a "spacer" is created. The latter provides for a certain distance between the growing peptide chain and the matrix of the resin and thereby decreases the unfavorable crowding around the polymer-bound amine-component. The reactants have better access to a less encumbered nucleophile.

The aminomethyl-copolymer could be used for the preparation of peptide amides [2] as well via acidolytic fission of the C−N bond in the benzylamine moiety:

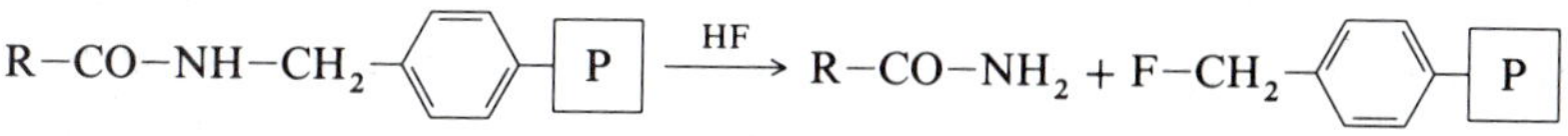

$$R-CO-NH-CH_2-\langle\rangle-[P] \xrightarrow{HF} R-CO-NH_2 + F-CH_2-\langle\rangle-[P]$$

For practical purposes, however, it is necessary to enhance the acid-sensitivity of this bond by the presence of a second benzene ring at the benzylic carbon atom. Benzhydrylamine resins indeed proved themselves in numerous syntheses of peptide amides

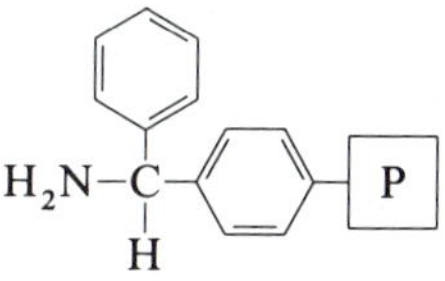

Polymers with even more acid-labile handles such as in

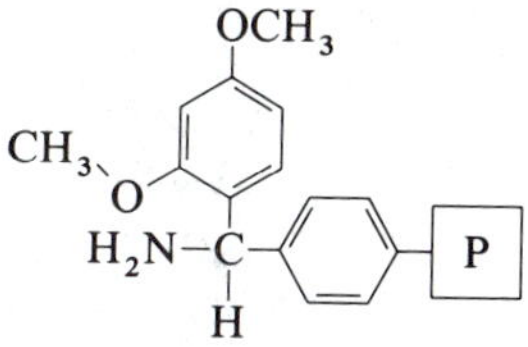

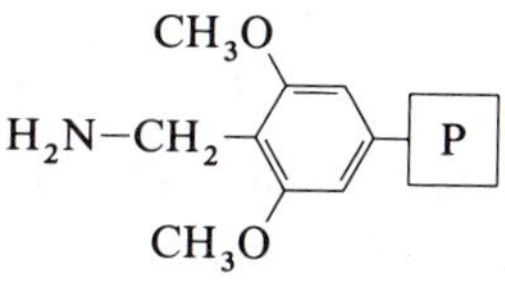

might turn out to have practical value.

---

[2] Quite a few peptide hormones have an amino acid amide rather than a free carboxyl group at their C-termini. Such peptide amides can be secured by ammonolysis of the anchoring ester bond

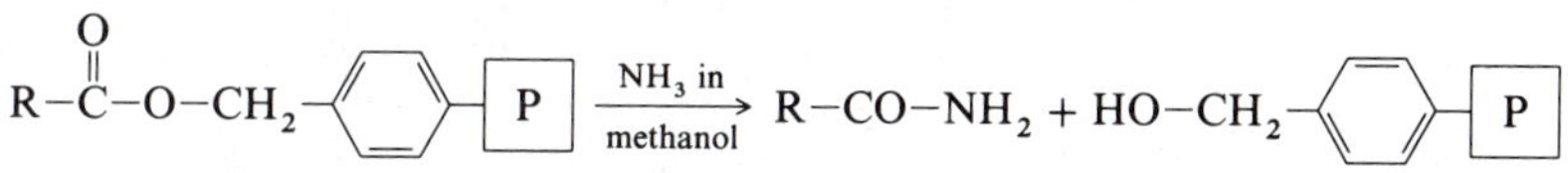

$$R-\overset{O}{\overset{\|}{C}}-O-CH_2-\langle\rangle-[P] \xrightarrow[methanol]{NH_3 \text{ in}} R-CO-NH_2 + HO-CH_2-\langle\rangle-[P]$$

but this reaction is accompanied by methanolysis. The resulting methyl esters are gradually converted to the desired amide

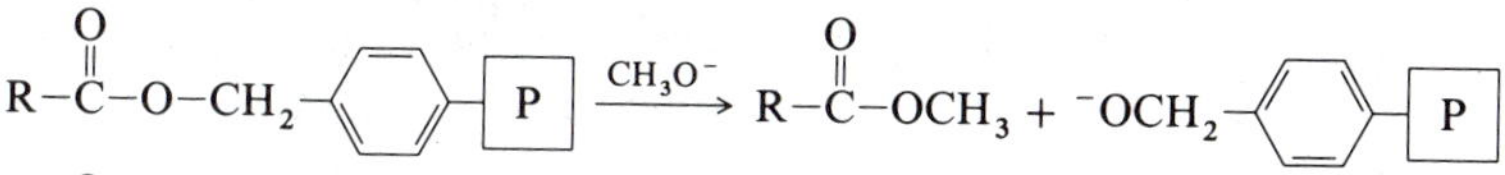

$$R-\overset{O}{\overset{\|}{C}}-O-CH_2-\langle\rangle-[P] \xrightarrow{CH_3O^-} R-\overset{O}{\overset{\|}{C}}-OCH_3 + {}^-OCH_2-\langle\rangle-[P]$$

$$R-\overset{O}{\overset{\|}{C}}-OCH_3 \xrightarrow{NH_3} R-CONH_2 + CH_3OH$$

except that the second reaction might require considerable time for completion. With tertiary amines used for the generation of the methoxide anion, methyl esters can be obtained as the sole products.

Finally, the styrene-divinylbenzene copolymer is certainly not the only starting material suitable for functionalization. Impermeable polymers were grafted and porous glass beads coated with organic polymers for derivatization, but these pioneering efforts had few followers so far. More auspicious seems to be, at this time, the application of *polyamides,* such as polyacrylamide, poly-dimethylacrylamide or polydimethylacrylamide-co-Boc-$\beta$-alanylacroyl-hexa-methylenediamine. Their promise stems from the compatibility of the support with the peptide chain.

The foregoing paragraphs give an oversimplified picture of insoluble supports in solid phase peptide synthesis. A detailed discussion of the physical chemistry of these materials, problems of diffusion, swelling in various solvents require a lengthy treatment that can not be fitted into the framework of a brief textbook. The interested reader must be referred to a comprehensive review article by Barany and Merrifield (1980).

## B. Removal of Temporary Blocking Groups

Following the anchoring of the blocked C-terminal amino acid to the insoluble polymeric support its amine-protecting group must be removed to allow the incorporation of the next amino acid, the penultimate residue in the sequence. For a considerable period of time acidolysis was used almost exclusively for this purpose. Accordingly blocking groups with pronounced sensitivity towards acids had to be selected and the tert.butyloxycarbonyl (Boc) group, applied from the earliest days of solid phase peptide synthesis, remained in a position of near-monopoly. From time to time more acid labile blocking groups such as the biphenylylisopropyloxycarbonyl (Bpoc), the phenylisopropyloxycarbonyl (Poc) or the 3,5-dimethoxyphenylisopropyloxycarbonyl (Ddz) groups

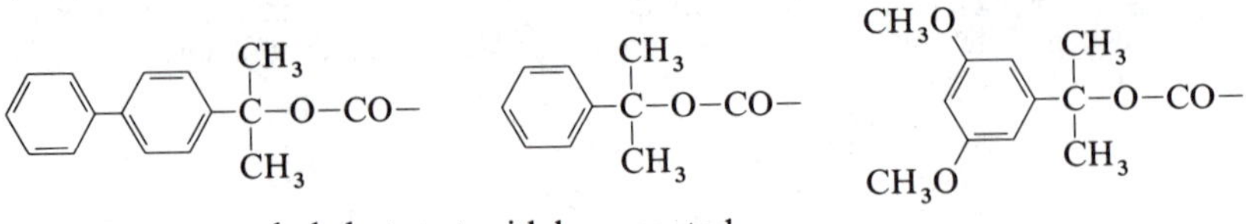

were recommended, but not widely accepted.

Hydrochloric acid in acetic acid, used for deblocking initially, was gradually replaced by trifluoroacetic acid. Since the commonly applied polymers do not swell enough in neat trifluoroacetic acid, it was necessary to dilute the reagent with a suitable solvent, usually dichloromethane. A minor, but not negligible side reaction, alkylation of sensitive side chains by the tert.butyl trifluoro-acetate generated in the reaction must be kept in mind. Yet, alternative acidic reagents, such as HCl in formic acid, methanesulfonic acid, *p*-toluenesulfonic acid in dioxane or 2-mercapto-ethanesulfonic acid, which serves also as scavenger in alkylation reactions, failed to gain popularity.

The probably most important alternative to the Boc group and acidolysis is the use of base-sensitive amine protecting groups. Application of the o-ni-trobenzenesulfenyl (Nps) group, sensitive to thiols, was repeatedly attempted but the practitioners seemed to be discouraged by the complications associated with the Nps approach. The limited shelf life of Nps amino acids necessitates their storage in the form of salts, which have to be converted to the free acid prior to coupling. Also, the masking group might prematurely be removed by 1-hydroxybenzotriazole, often present in the coupling mixture. Adaptation of the 9-fluorenylmethyloxycarbonyl (Fmoc) group for solid phase peptide synthesis appears to be much more successful. The reagent of choice in the deblocking step is a dilute solution of piperidine in dimethylformamide (DMF). To achieve complete deprotection of the amino group the reaction is allowed to proceed at room temperature for a few minutes and the cleavage step is then repeated for a somewhat longer period of time. The byproducts of the reaction, dibenzofulvene and the tertiary amine formed through the addition of piperidine to its double bond

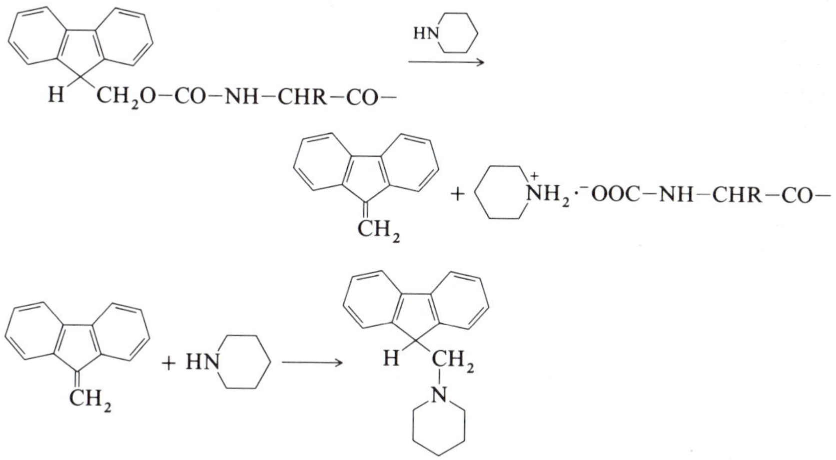

$$\text{H—CH}_2\text{O—CO—NH—CHR—CO—} \quad \xrightarrow{\text{HN}} $$

$$\text{(fluorenylidene)}=\text{CH}_2 \;+\; \text{piperidinium}\,\overset{+}{\text{NH}}_2, \;^-\text{OOC—NH—CHR—CO—}$$

$$\text{(fluorenylidene)}=\text{CH}_2 \;+\; \text{HN(piperidine)} \longrightarrow \text{H—CH}_2\text{—N(piperidine) (fluorenyl)}$$

are readily washed out with DMF. The piperidinium salt of the carbamoic acid produced in the reaction decomposes spontaneously to yield piperidine, carbon dioxide and the *free amine:*

$$\overset{+}{\text{NH}}_2, \;^-\text{OOC—NH—CHR—CO—} \longrightarrow \text{NH (piperidine)} + CO_2 + H_2\text{N—CHR—CO—}$$

Thus, unlike in deprotection of acid labile blocking groups by acidolysis, no conversion of salts to free amines is needed and the nucleophile is ready, as formed, for the next coupling reaction. This represents a major simplification of the procedure, a shorter cycle. The number of washing steps is further reduced by the circumstance that both coupling and deprotection are carried out in the same solvent, DMF. The principal advantage of temporary blocking by the Fmoc group is, however, the opportunity to use semipermanent masking groups based on formation of the tert.butyl cation. This orthogonal combination eliminates several shortcomings of the earlier favored Merrifield resin — Boc protection approach. For instance the risk of alkylation and the need of drastic acidolytic conditions in the final deprotection-separation from the resin are avoided. Of course a modified polymeric support is required with an anchoring bond that can be cleaved under mild conditions.

## C. Coupling: Acylation of the Resin-bound Amine-component

On reviewing the coupling methods proposed for application in solid phase peptide synthesis one finds that most known methods of peptide bond formation have been tried but no new approach was designed specially for the acylation of the polymer bound amine component. Therefore we will limit the discussion in this section to two well established procedures, coupling with the help of carbodiimides and acylation via active esters.

Right from the introduction of the solid phase method dicyclohexylcarbodiimide (DCC) was regarded as the reagent of choice for coupling and it is still applied in the majority of syntheses. The ready availability of the reagent, at a low price, is only a minor reason for its popularity. A more important feature of the procedure is the rapid course of activation and coupling, although the high rates observed with both the amine- and the carboxyl-component in solution are not fully reproduced with polymer bound amine-components. The matrix of the resin acts both as diluent and as steric hindrance. Nevertheless, reaction rates remain satisfactory. This is an important consideration, because in the automated solid phase synthesis of long peptide chains sometimes half a dozen or more amino acid residues are incorporated in a 24 hour day. A further advantageous feature of acylation with DCC is the ease with which the N,N'-dicyclohexyl urea (DCU) byproduct is removed from the system. When coupling is carried out in solution DCU, which is poorly soluble in most organic solvents, is removed by filtration. In the solid phase technique one would expect its accumulation between and inside the resin particles. Dissolution of DCU in and washing with ethanol would remove the byproduct but, fortunately even this is unnecessary because the trifluoroacetic acid-dichloromethane mixture used for deprotection is a fairly good solvent of DCU. Some laboratories further improve this situation by applying diisopropylcarbodiimide rather than DCC, since N,N'-diisopropyl-urea is more soluble than DCU.

The classical way of coupling with the help of DCC (Sheehan and Hess 1955) is to add one mole of the reagent to a solution which contains equimolar amounts of the two components to be joined through a peptide bond. This *coupling reagent mode* is not best suited for solid phase work. Already during the formation of the O-acylisourea intermediate the latter is attacked by the amine-component, but even more readily by yet unreacted molecules of the carboxyl-component. This second pathway yields a symmetrical anhydride, a good acylating agent in its own right, leaves, however, half of the DCC unreacted.

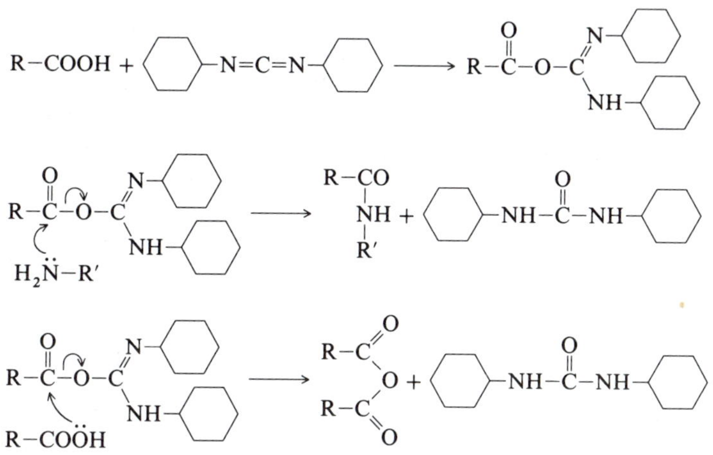

As the symmetrical anhydride acylates the amine-component, one equivalent of the carboxyl-component is set free and reacts, in turn, with the still available DCC. Thus both DCC and the O-acylisourea intermediate are completely used up, but only *gradually*. Yet the highly reactive O-acylisoureas if unattacked by an external nucleophile rearrange under intramolecular nucleophilic attack (cf. Chapter V) to stable N-acylureas and this is only one of the side reactions (cf. Chapter VII) which follow from long-lived overactivated intermediates. The coupling reagent-mode is, therefore far from ideal in syntheses carried out in solution and even less desirable in solid phase work where the amine-component is resin bound and hence less readily available. A technique, in which the free amino group of the peptidyl resin is first converted to a *salt* of the blocked amino acid to be incorporated and only then coupled with DCC gave relatively good results, probably because activation is immediately followed by the attack of the amino group held in proximity in the ion-pair.

A major improvement in DCC mediated coupling was brought about by the use of the reagent in "symmetrical-anhydride-mode". This means that *one* equivalent of the coupling reagent is added to the solution containing *two* equivalents of the blocked amino acid:

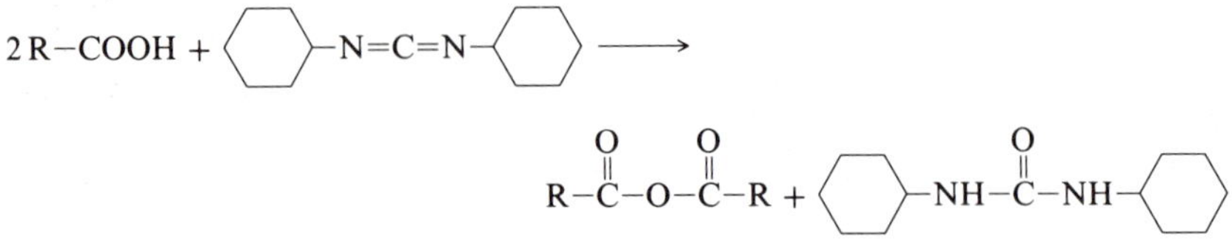

$$2R-COOH + \left\langle\ \right\rangle -N=C=N-\left\langle\ \right\rangle \longrightarrow$$

$$R-\overset{O}{\underset{\|}{C}}-O-\overset{O}{\underset{\|}{C}}-R + \left\langle\ \right\rangle-NH-\overset{O}{\underset{\|}{C}}-NH-\left\langle\ \right\rangle$$

Attack of the carboxyl-component on the O-acylisourea intermediate is facile, unimpeded by the polymer. In fact, the symmetrical anhydride can first be generated in solution and if desired DCU removed prior to the addition of the acylating mixture to the resin-bound amine-component. In this mode of coupling the lifetime of the overly reactive O-acylisourea is significantly reduced with the simultaneous reduction in the extent of side reactions such as N-acylurea formation. The symmetrical anhydride approach, considered earlier as too expensive, became with the decrease in price of Boc-amino acids quite popular in recent years.

At this point it should be noted that symmetrical anhydrides of benzyloxycarbonyl and tert.butyloxycarbonyl amino acids were isolated time and again in crystalline form but were not commercially available, probably because of their limited shelflife. The situation seems to be different with Fmoc amino acid symmetrical anhydrides

$$\text{H} \quad CH_2O-CO-NH-CHR-\overset{O}{\underset{\|}{C}}-O-\overset{O}{\underset{\|}{C}}-CHR-NH-CO-OCH_2\text{H}$$

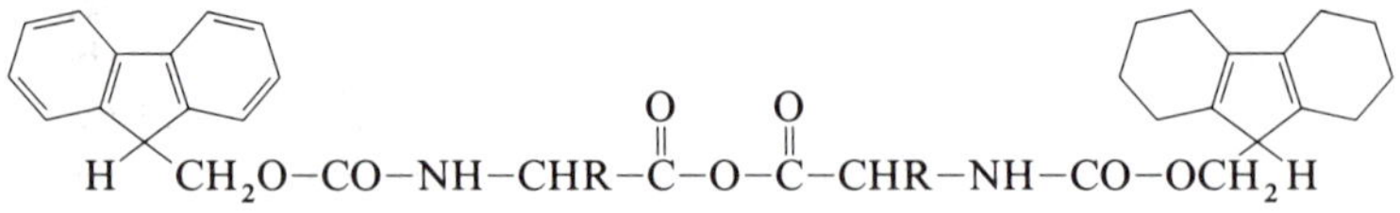

which appear to be sufficiently stable.

Lifetime of the overactivated intermediate in DCC mediated reactions can be limited also by the addition of the powerful auxiliary nucleophile 1-hydroxybenzotriazole (HOBt). The excellent acylation properties of the active esters thus generated

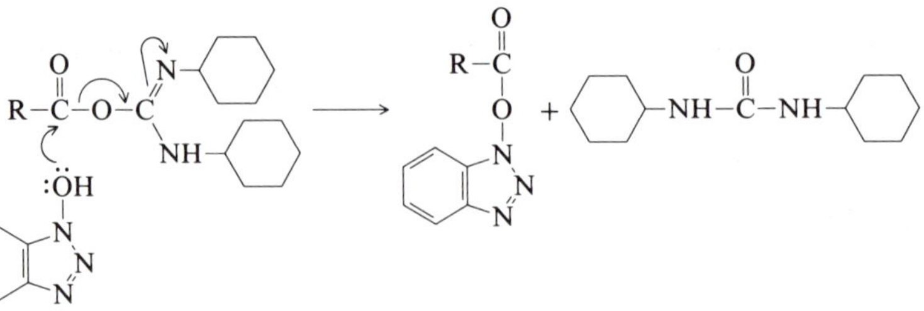

$$R-\overset{O}{\underset{\|}{C}}-O-\overset{N-\langle\ \rangle}{\underset{NH-\langle\ \rangle}{C}} \longrightarrow \overset{O}{\underset{\|}{C}} + \left\langle\ \right\rangle-NH-\overset{O}{\underset{\|}{C}}-NH-\left\langle\ \right\rangle$$

have been discussed in Chapter V.

Isolated, stable active esters have been used in solid phase peptide synthesis, but the improvements achieved with respect to homogeneity of the products were counterbalanced by the long time required for completion of the acylation reaction. Only the p-nitrophenyl esters of Boc-asparagine and glutamine found broad acceptance. With their help the noxious dehydration reaction which occurs during activation with DCC could be avoided. In acylation with the purified active esters

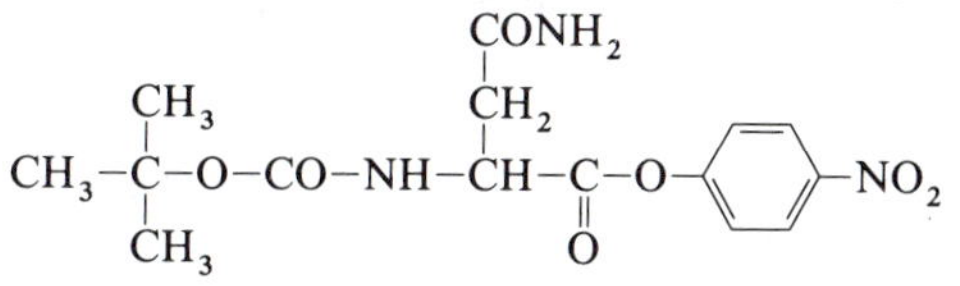

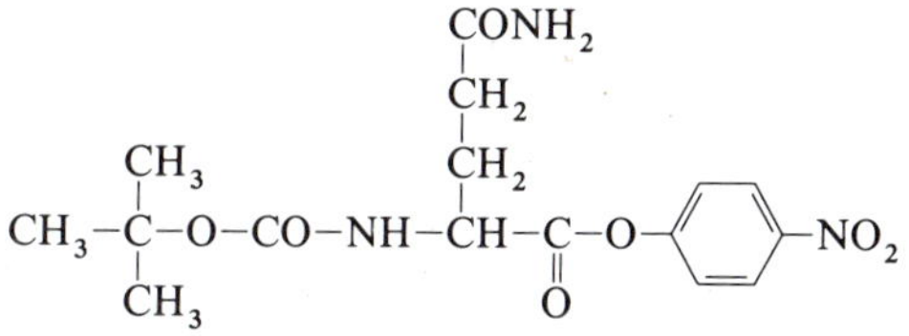

$$
CH_3-\underset{\underset{CH_3}{|}}{\overset{\overset{CH_3}{|}}{C}}-O-CO-NH-\underset{}{\overset{\overset{CONH_2}{|}}{\underset{\underset{O}{\|}}{\overset{\overset{CH_2}{|}}{CH}}}-\underset{}{C}-O-\langle\!\!\bigcirc\!\!\rangle-NO_2
$$

$$
CH_3-\underset{\underset{CH_3}{|}}{\overset{\overset{CH_3}{|}}{C}}-O-CO-NH-\underset{\underset{O}{\|}}{\overset{\overset{\overset{\overset{CONH_2}{|}}{CH_2}}{|}}{\underset{}{\overset{CH_2}{|}}{CH}}}-C-O-\langle\!\!\bigcirc\!\!\rangle-NO_2
$$

no incorporation of β-cyanoalanine or γ-cyano-α-aminobutyric acid takes place.

Better reaction rates were observed with the already inherently more reactive o-nitrophenyl esters, in part because they are less sensitive to steric hindrance than their para-substituted analogs. Bulkyness of the activating groups renders N-hydroxysuccinimide esters and pentachlorophenyl esters less well suited for acylation inside the matrix of a polymer than in solution. Only the introduction of the powerful pentafluorophenyl esters led to a revival of the active ester idea in the praxis of solid phase peptide synthesis. Pentafluorophenyl esters of Fmoc amino acids

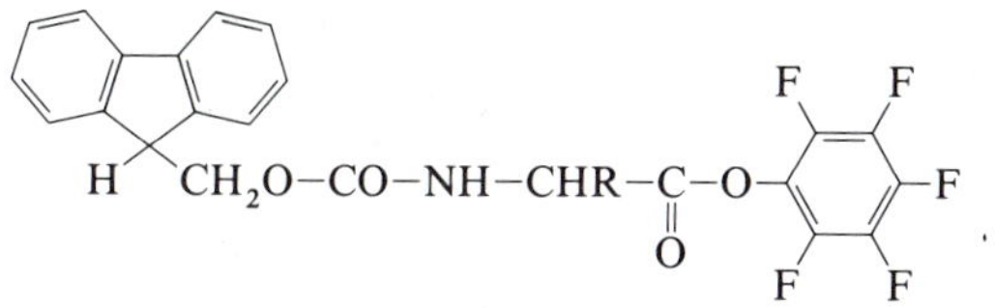

$$
\text{H}-\text{CH}_2\text{O}-\text{CO}-\text{NH}-\text{CHR}-\underset{\underset{O}{\|}}{C}-O-\langle\!\!\bigcirc\!\!\rangle_{F}^{F}\cdots F
$$

are now commercially available and frequently used. Similarly effective are the esters of 3-hydroxy-3,4-dihydro-1,2,3-benzotriazine-4-one

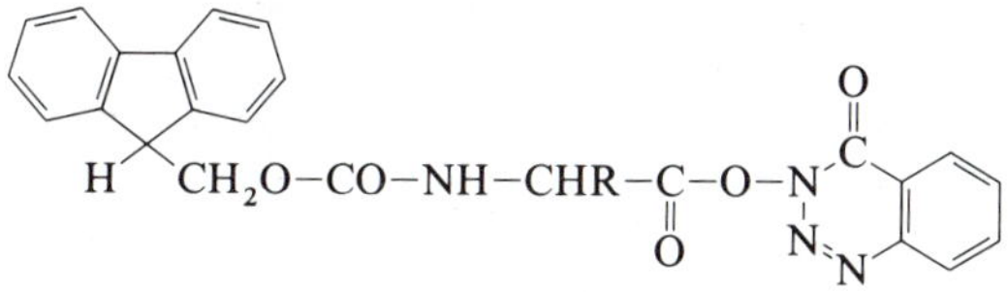

$$
\text{H}-\text{CH}_2\text{O}-\text{CO}-\text{NH}-\text{CHR}-\underset{\underset{O}{\|}}{C}-O-N
$$

Practical acylation rates can be achieved also with the less reactive p-nitrophenyl esters through catalysis with 3-hydroxy-3,4-dihydroquinazoline-4-one in dimethylformamide, particularly in the presence of a tertiary amine.

## D. Separation of the Peptide from the Polymeric Support and Final Deprotection

The two steps mentioned in the title of this section are not necessarily carried out in the same operation. For instance, a completed segment of a long chain can be separated from the support, still in blocked form, by hydrazinolysis

$$-NH-CHR-CO-OCH_2 \overset{H_2NNH_2}{\longrightarrow}$$

$$-NH-CHR-CO-NHNH_2 + HOCH_2$$

and the peptide hydrazide used, after conversion to the azide, for acylation of the amino group of a second segment. Similarly, when peptide amides are prepared through ammonolysis of the anchoring ester bond, fission of the link between peptide and resin takes place without general deblocking. In the majority of syntheses, however, acidolysis is applied both for cleavage of the anchoring bond and for the removal of semipermanent blocking groups from side chain functions. An acid labile masking group at the N-terminal α-amino group allows complete deprotection in a single operation.

The most frequently used combination of blocking groups, Boc for transient protection of the α-amine function and benzyl-derived or other relatively acid resistant groups for the semipermanent blocking of side chain functions, together with a benzyl ester type anchoring bond, leaves few choices at the concluding step. Hydrogenolysis has been considered and can be implemented but not always with full success. Impregnation of the resin with a solution of palladium acetate followed by hydrogenation yields a fine dispersion of the metal inside the gel, hence catalytic reduction becomes possible. The rate of the reaction, however, is far from satisfactory and the process has to be executed at high pressure and elevated temperature. Thus, for the crucial operation of final deblocking the practitioner must chose between various acidic reagents, all very strong acids.

In the earliest period of the development of solid phase peptide synthesis hydrobromic acid in trifluoroacetic acid was applied, often with good results. Some side chain blocking groups, such as the nitro group or the *p*-toluenesulfonyl group used for masking of the guanidine function or the S-benzyl group in cysteine side chains are, however, fairly resistant to the reagent, probably because HBr is only moderately soluble in trifluoroacetic acid and even continued introduction of the gas fails to provide HBr concentrations comparable to those reached in acetic acid. (Unfortunately, the latter is unsuited because deprotection with HBr in acetic acid is accompanied by partial O-acetylation of the serine side chain.) The application of liquid hydrogen fluoride as acidic reagent (Sakakibara and Shimonishi 1965) revolutionized the solid phase technique. The gas is condensed in a teflon container which already contains the peptidyl resin. Liquid HF penetrates the particles of the polymeric material, cleaves the bond between peptide and polymer and dissolves the released peptide. Benzyl esters, benzyl ethers and side chain benzyloxycarbonyl groups are removed at the same time, usually in less than 20 minutes at 0 °C. Under these conditions benzyl groups substitued with electronwithdrawing atoms or groups are incompletely cleaved and the same is true for other relatively acid stable blocking groups. Thus, for the removal of nitro groups or p-toluenesulfonyl groups from arginine side chains and S-benzyl groups from cysteine side chains unmasking must be carried out at room temperature and the reaction time extended for one hour. Under these conditions an important side reaction must be reckoned with, to wit, the well known $N \rightarrow O$ acyl-migration at serine residues:

$$-NH-CHR-CO-NH-\overset{\overset{\displaystyle HO-CH_2}{|}}{CH}-CO- \underset{HO^-}{\overset{H^+}{\rightleftharpoons}} \quad \overset{\overset{\displaystyle -NH-CHR-CO-O-CH_2}{}}{\underset{H_3N-\overset{+}{C}H-CO-}{}}$$

This unwanted rearrangement can be reversed by treatment of the product with sodium bicarbonate in water, but not without some accompanying hydrolysis. Several other shortcomings of the method are known, such as the HF catalyzed Friedel-Crafts reaction between $\gamma$-carboxyl groups in glutamyl residues and aromatic nuclei in the polymer:

$$\overset{\overset{\displaystyle CH_2-COOH}{|}}{\underset{\underset{\displaystyle -NH-CH-CO-}{|}}{CH_2}} + \langle\!\!\!\bigcirc\!\!\!\rangle\!-\!\boxed{P} \overset{HF}{\longrightarrow} \overset{\overset{\displaystyle CH_2-\overset{\overset{\textstyle O}{\|}}{C}-\langle\!\!\!\bigcirc\!\!\!\rangle\!-\!\boxed{P}}{}}{\underset{\underset{\displaystyle -NH-CH-CO-}{|}}{CH_2}}$$

This side reaction is relatively innocuous because the by-product is irreversibly bound to the polymer and only the yield is affected not the purity of the synthetic peptide. More disturbing is the succinimide ring formation at aspartyl residues exposed to HF. Alkylation of the indole ring in tryptophan, the phenolic side chain in tyrosine and the sulfur atom in methionine must be suppressed by the addition of scavengers. The often applied anisole is less than unequivocal in this role: it can be the source of methyl groups which convert the methionine thioether to a tertiary sulfonium derivative. The acid stable thioanisole seems to be a better scavenger.

Some improvements were noted in a two step-cleavage, in which first HF diluted with dimethyl sulfide is used, followed by the application of neat hydrogen fluoride.

In spite of these shortcomings liquid hydrogen fluoride is clearly an efficient reagent for the concluding step of solid phase peptide synthesis: it proved itself useful in many laboratories. Of course, in addition to the care required while working with this dangerous material, attention must be paid to the possible by-products in the material separated from the insoluble support. Fortunately, with advanced methods of chromatography it became possible to secure pure peptides even from fairly complex mixtures. Nonetheless, HF is probably not the perfect reagent for the synthesis of delicate peptide sequences. Application of a solution of trifluoromethanesulfonic acid in trifluoroacetic acid in the presence of cresols is more convenient since no special apparatus is needed: the reaction can be carried out in glass equipment. Yet, with this powerful acidic reagent one should be prepared to observe at least some of the side reactions noted in the use of liquid hydrogen fluoride. It should also be remembered that acidolysis results in amine *salts*. Hence the peptides recovered after separation from the polymeric support are hydrogen fluorides, trifluoromethanesulfonates, etc. If acetates or hydrochlorides appear more desirable, then fission from the resin must be followed by treatment with an appropriate ion-exchange material.

The already discussed orthogonal combination of the 9-fluorenylmethyloxycarbonyl (Fmoc) group for the transient protection of α-amino functions with tert.butyl cation forming groups for the semipermanent blocking of side chain functions, is quite auspicious for solid phase peptide synthesis. With acid labile anchoring based on *p*-alkoxy-substituted benzyl esters or on benzhydrylamines, final deprotection and fission of the peptide-resin bond can be achieved in a single step by acidolysis with trifluoroacetic acid. For satisfactory swelling of the polymer an additional solvent, such as dichloromethane, might be needed and it is advisable to use scavengers to suppress tert.butylation caused by tert.butyl trifluoroacetate generated in the deblocking reaction. The Fmoc-tert.butyl combination represents a major improvement over non-orthogonal approaches which entail drastic acidolysis in the concluding step.

## E. Some Problems Encountered in Solid Phase Peptide Synthesis

The side reactions and imperfections discussed in this section are not particular or limited to the solid phase procedure. They do occur in syntheses performed in solution, but rather infrequently. The matrix of the polymeric support hinders both acylation and deprotection and if these reactions are prematurely terminated, by-products will form which might amount to not negligible impurities in the synthetic material. The analogous by-products formed in solution are readily recognized and can be eliminated when the intermediates are isolated and purified. This possibility is practically absent in solid phase work, thus it is imperative to address these problems.

## 1. Premature Chain-termination

Deblocking with HCl in acetic acid or with trifluoroacetic acid can leave, even after thorough washing, small amounts of acetic acid or trifluoroacetic acid attached to the peptidyl-resin. The carboxyl groups of these acids are activated in the next coupling step, certainly by carbodiimides, but potentially also by symmetrical anhydrides or active esters

$$R-\overset{O}{\underset{\parallel}{C}}-O-\overset{O}{\underset{\parallel}{C}}-R + CH_3COOH \overset{base}{\rightleftharpoons} CH_3-\overset{O}{\underset{\parallel}{C}}-O-\overset{O}{\underset{\parallel}{C}}-R + R-COOH$$

$$R-\overset{O}{\underset{\parallel}{C}}-O-\langle\bigcirc\rangle-NO_2 + CH_3COOH \overset{base}{\rightleftharpoons} CH_3-\overset{O}{\underset{\parallel}{C}}-O-\langle\bigcirc\rangle-NO_2 + R-COOH$$

$$R-\overset{O}{\underset{\parallel}{C}}-O-\langle\bigcirc\rangle-NO_2 + CH_3COOH \overset{base}{\rightleftharpoons} R-\overset{O}{\underset{\parallel}{C}}-O-\overset{O}{\underset{\parallel}{C}}-CH_3 + HO-\langle\bigcirc\rangle-NO_2$$

and the activated species will acylate a fraction of the amine component. Trifluoroacetyl (and also formyl) peptides can be unblocked by brief treatment with a base such as aqueous piperidine, but acetylation is irreversible. The obvious remedy is to insist on the complete removal of carboxylic acids from the peptidyl-polymer prior to coupling.

A second type of chain-termination is caused by ketones and aldehydes that might contaminate the solvents. For instance dichloromethane often contains formaldehyde. Such carbonyl compounds react with the N-terminal dipeptide portion of a chain to give imidazolidinones:

$$\underset{H}{\overset{H}{\diagdown}}C\underset{HN}{\overset{H_2N-CHR}{\diagup}}\overset{\diagdown}{\underset{\diagup}{C=O}}C=O \longrightarrow \underset{H}{\overset{H}{\diagdown}}C\underset{N}{\overset{HN-CHR}{\diagup}}\overset{\diagdown}{\underset{\diagup}{C=O}}C=O \quad + H_2O$$
$$CHR'-CO-NH-CHR''-CO- \qquad\qquad CHR'-CO-NH-CHR''-CO-$$

In the condensation products the terminal amino group of the peptide is masked and thus unavailable for acylation. Careful purification of solvents prevents this undesirable side reaction.

The well known cyclization of N-terminal glutamine residues

$$\overset{O}{\diagdown}\underset{H_2N}{\overset{\diagup}{C}}\underset{H_2\ddot{N}-CH-CO-NH-CHR-CO-}{\overset{CH_2}{\diagdown}CH_2} \overset{-NH_3}{\longrightarrow} O=C\underset{NH-CH-CO-NH-CHR-CO-}{\overset{CH_2}{\diagdown}CH_2}$$

to pyroglutamyl peptides is catalyzed by weak acids. Of course, the chain can not be further lengthened unless the ring is opened by treatment with ammonia or the pyroglutamyl residue eliminated with the help of a specific enzyme.

## 2. Deletion

Independently from the technique followed both deblocking and acylation must be carried to completion. In solid phase synthesis it is really mandatory to unmask even the last traces of the terminal amino group. If less than the entire amount of the amine-component is available for acylation then in the subsequent incorporation of the next amino acid residue two blocked peptides are produced:

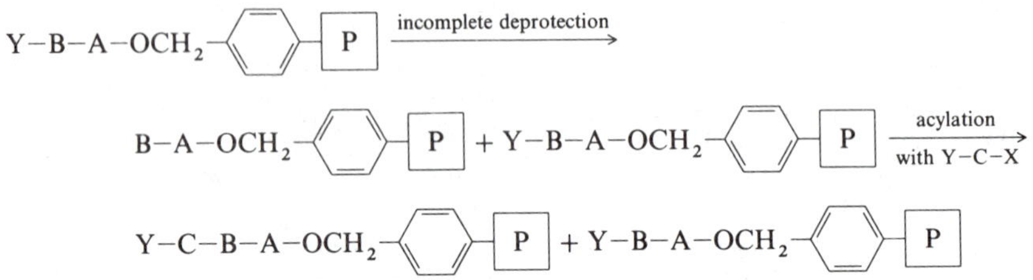

Y−B−A−OCH$_2$−⟨ ⟩−P  $\xrightarrow{\text{incomplete deprotection}}$

B−A−OCH$_2$−⟨ ⟩−P + Y−B−A−OCH$_2$−⟨ ⟩−P  $\xrightarrow[\text{with Y−C−X}]{\text{acylation}}$

Y−C−B−A−OCH$_2$−⟨ ⟩−P + Y−B−A−OCH$_2$−⟨ ⟩−P

(Y is a temporary blocking group)

In the following cycle the Y-group is eliminated and the protected residue D is incorporated. This way two blocked peptide intermediates are generated:

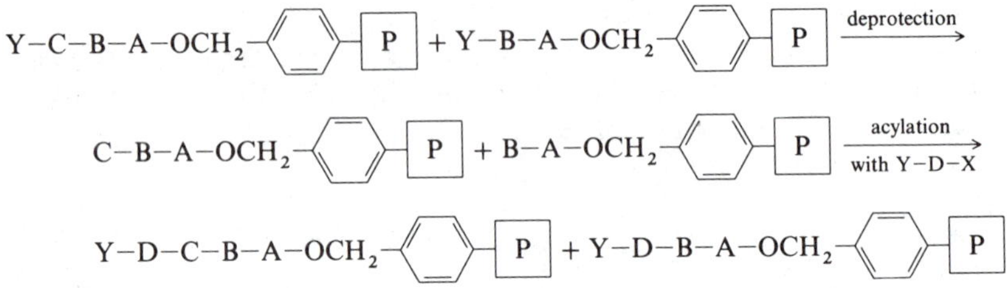

Y−C−B−A−OCH$_2$−⟨ ⟩−P + Y−B−A−OCH$_2$−⟨ ⟩−P  $\xrightarrow{\text{deprotection}}$

C−B−A−OCH$_2$−⟨ ⟩−P + B−A−OCH$_2$−⟨ ⟩−P  $\xrightarrow[\text{with Y−D−X}]{\text{acylation}}$

Y−D−C−B−A−OCH$_2$−⟨ ⟩−P + Y−D−B−A−OCH$_2$−⟨ ⟩−P

(Y is a temporary blocking group)

The second of these lacks residue C; it is a "deletion sequence". The picture is further complicated if incomplete deprotection occurs in several or, a forbidding possibility, in all cycles of the chain-building process. Therefore one has to insist on complete deblocking even at the expense of prolonged reaction time. Because of the risk of side reactions, such as alkylation, during the unmasking reaction, it might be preferable to use first a rather brief deblocking period and to eliminate the harmful materials generated in the process. Then deprotection is repeated, this time for a more extended period.

The consequences of incomplete acylation are quite similar. If a fraction of the resin-bound dipeptide B−A is left unacylated during the incorporation of residue C, then after the following deprotection and chain lengthening with (blocked) residue D, the intermediate Y-D-C-B-A will be contaminated with Y-D-B-A, a "deletion peptide". This kind of complexity is particularly likely to occur in acylation with derivatives of valine and isoleucine. Some amino groups seem to lie in secluded areas of the polymer matrix and are further encumbered by the growing peptide chain itself. Such amino groups are unaccessible for hindered acylating agents and remain unchanged even at equilibrium. A partial remedy against the formation of deletion sequences can be found in the intentional termination of the chains with unreactive N-termini. Unhindered acylating reagents such as acetic anhydride, p-nitrophenyl acetate or acetylimidazole can reach the hidden amino groups. The thus terminated ("truncated") chains are much more different from the target compound than the various deletion sequences and hence more easily separated from it. Several sophisiticated acylating agents have been proposed to facilitate the isolation of blocked chains from the principal product. Colored acyl groups provide simple visual control in the separation of peptides, groups with ionic properties allow the elimination of the blocked chains with the help of ion-exchangers. In the general practice of solid phase peptide synthesis only acetylation is applied.

A special case of deletion is the simultaneous loss of two residues, the amino acid attached via an ester bond to the polymer together with the penultimate residue. The known tendency of dipeptide esters for cyclization is responsible for this side reaction. Elimination of the two residues as a diketopiperazine

leaves the styrene-divinylbenzene-copolymer in the form of the hydroxymethyl derivative that is acylated during the incorporation of the third residue. As a consequence a new chain is started from which the C-terminal amino acid and the preceding residue are missing. This usually only minor side reaction is more pronounced when proline or glycine are involved. Diketopiperazine formation can be kept at an acceptable level by avoiding delays after deprotection of the dipeptidyl resin and by creating favorable conditions for the incorporation of the third residue. The latter should be applied in the form of a highly reactive derivative and in high concentration.

It is difficult to anticipate all possible imperfections of the chain-building process and even more problematic to find perfect remedies for undesired side reactions. Therefore, one has to resort to monitoring the completion of the individual steps. Titration of unblocked amino groups with a strong acid, for instance perchloric acid, has been proposed as a method by which the process of deprotection and the progress of acylation can be followed. In acylation with active esters the absorption of the eliminated material (e.g. p-nitrophenol) in the ultraviolet can be used for monitoring. Esters of 3-hydroxy-3,4-dihydro-1,2,3-

benzotriazine give rise to a yellow color due to the enolate salt of the amine. Once acylation of the latter is complete the color of the resin disappears. In most laboratories, however, the end point is established by treating a small sample of the peptidyl polymer with ninhydrin. A positive test prompts repetition of the coupling step. It seems to us that solid phase peptide synthesis is fully automated only if the procedure includes analytical control of both deprotection and coupling. At this time the commercially available instruments are not geared to do that. On the other hand the automatic sequencers easily detect the presence of "deletion sequences" in the crude synthetic material: a "preview" of an amino acid, that is its appearence one cycle earlier than expected, is clear indication for the deletion of the preceding residue from the sequence.

# References

Bodanszky, M., du Vigneaud, V.: J. Amer. Chem. Soc. *81*, 5688 (1959)
Letsinger, R. L., Cornet, M. J.: J. Amer. Chem. Soc. *85*, 3045 (1963)
Merrifield, R. B.: J. Amer. Chem. Soc. *85*, 2149 (1963)
Merrifield, R. B.: J. Amer. Chem. Soc. *86*, 304 (1964)
Sheehan, J. C., Hess, G. P.: J. Amer. Chem. Soc. *77*, 1067 (1955)
Sakakibara, S., Shimonishi, Y.: Bull. Chem. Soc. Jpn. *38*, 1412 (1965)

# Additional Sources

Barany, G., Merrifield, R. B.: Solid Phase Peptide Synthesis, in The Peptides Vol. 2, Gross, E., Meienhofer, J., eds. pp. 1–284, Academic Press, New York 1980
Birr, C.: Aspects of the Merrifield Peptide Synthesis, Springer Verlag, Berlin 1978
Stewart, J. M., Young, J. D.: Solid Phase Peptide Synthesis, 2nd ed., Pierce Chemical Co., Rockford, Illinois 1984

# XI. Methods of Facilitation

Solid phase peptide synthesis is the only widely practiced technique of facilitation but several other methods have also been proposed for the rapid construction of long peptide chains. Some of these alternative approaches have certain advantages over solid phase synthesis: reactions carried out in solution are not affected by the rate-limiting control of the gel. Also, where isolation of intermediates is possible, these can be analyzed and purified. A major disadvantage, however, common to the various techniques discussed below, is that they are less conducive to mechanization and automation than the Merrifield method.

## A. Chains Attached to a Soluble Polymer (The Liquid Phase Method)

An amino acid attached to a soluble polymer through its carboxyl group can be acylated by an appropriately blocked and activated amino acid and synthesis of a peptide chain can be started in this manner. The intermediate peptidyl-polymers are precipitated and washed to remove unreacted starting materials and by-products, the purified material redissolved, deblocked, acylated in solution and so on. In the first realization of this principle (Shemyakin et al. 1965) high molecular weight polystyrene (ca. 200.000 daltons) was applied, of course without crosslinking. The aminoacyl- and peptidyl-polymers are soluble in tetrahydrofurane, dimethylformamide, dimethylsulfoxide and other polar solvents. After completion of the necessary reactions the intermediates are separated from solution by dilution with water, the precipitate washed and the process continued until the desired peptide chain has been assembled. Lower molecular weight polystyrene (ca. 20.000 daltons) was used in the same manner except that gel-filtration rather than precipitation had to be used for the separation of the peptidyl-polymer intermediates. The best developed technique, however, is based on a different kind of polymer, namely polyethyleneglycol (PEG)

$$HO-CH_2-CH_2-(O-CH_2-CH_2-)_nO-CH_2-CH_2-OH$$

or, if a monofunctional supporting material is preferred, on its monoethyl ether

$$HO-CH_2-CH_2-(O-CH_2-CH_2-)_nO-CH_2-CH_2-OCH_3$$

The first amino acid is linked to PEG via an "anchoring" group or handle:

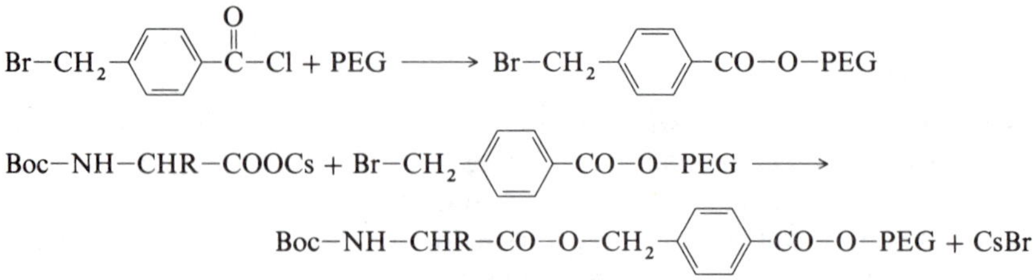

$$Br-CH_2-\langle\bigcirc\rangle-\overset{\overset{\displaystyle O}{\|}}{C}-Cl + PEG \longrightarrow Br-CH_2-\langle\bigcirc\rangle-CO-O-PEG$$

$$Boc-NH-CHR-COOCs + Br-CH_2-\langle\bigcirc\rangle-CO-O-PEG \longrightarrow$$

$$Boc-NH-CHR-CO-O-CH_2-\langle\bigcirc\rangle-CO-O-PEG + CsBr$$

The solubilizing effect of PEG permits coupling in aqueous solution as well. In this case a water soluble carbodiimide, such as

$$CH_3CH_2-N=C=N=CH_2CH_2CH_2N(CH_3)_2.HCl$$

can mediate the reaction.

With target peptides of moderate size soluble polymers provide a practical technique of facilitation that yields products in satisfactory yield and homogeneity. The method, however, has some inherent limitations as well. With increasing chain length the physical properties of the intermediate peptidyl-polymers gradually change from the properties of PEG to the properties of the blocked peptide chain. The solubility related problems associated with many high molecular weight blocked peptides have already been discussed in previous chapters.

## B. The Picolyl Ester Method

The peptide-research group at Oxford University, led by G. T. Young proposed an interesting and important combination of solid phase peptide synthesis and synthesis in solution (Camble et al. 1968). They provided the C-terminal amino acid with a semipermanent carboxyl-protecting group that generates a cation on protonation. Therefore, the blocked intermediates of the syntheses are readily bound to acidic cation-exchange resins and can thus be separated from excess starting materials and by-products. The reactions of coupling and deprotection are carried out in solution by well established methods.

The semipermanent blocking selected for this purpose is protection of the C-terminal carboxyl group in the form of picolyl esters. 4-Hydroxymethyl-pyridine is acylated with an acylamino acid, usually with the help of a carbodi-imide:

$$CH_3-\underset{\underset{CH_3}{|}}{\overset{\overset{CH_3}{|}}{C}}-O-CO-NH-CHR-COOH + HOCH_2-\langle\!\!\!\bigcirc\!\!\!N + $$

$$\bigcirc-N=C=N-\bigcirc \longrightarrow$$

$$CH_3-\underset{\underset{CH_3}{|}}{\overset{\overset{CH_3}{|}}{C}}-O-CO-NH-CHR-CO-OCH_2-\langle\!\!\!\bigcirc\!\!\!N + \bigcirc-NH-\overset{O}{\overset{||}{C}}-NH-\bigcirc$$

The picolyl ester group is quite resistant to acids, because protonation of the pyridine ring destabilizes the potential carbo-cation at the benzylic carbon atom. Hence, the tert.butyloxycarbonyl group, applied for the blocking of α-amino functions, can be selectively cleaved by trifluoroacetic acid and the resulting amine salt used[1] in the following coupling step:

$$CH_3-\underset{\underset{CH_3}{|}}{\overset{\overset{CH_3}{|}}{C}}-O-CO-NH-CHR-CO-OCH_2-\langle\!\!\!\bigcirc\!\!\!N \xrightarrow{CF_3COOH}$$

$$CF_3COO^- \cdot H_3\overset{+}{N}-CHR-CO-OCH_2-\langle\!\!\!\bigcirc\!\!\!N \xrightarrow[\text{Boc-NH-CHR'-COOH}]{N(C_2H_5)_3;\ DCC/HOBt}$$

$$CH_3-\underset{\underset{CH_3}{|}}{\overset{\overset{CH_3}{|}}{C}}-O-CO-NH-CHR'-CO-NH-CHR-CO-OCH_2-\langle\!\!\!\bigcirc\!\!\!N$$

The blocked dipeptide derivative is then adsorbed on a cation-exchange resin. The initially recommended sulfoethyl-sephadex C-25, applied in hydrogen cycle, was replaced later by a macroreticular resin, Amberlyst-15 saturated with 3-bromopyridine to avoid premature cleavage of acid-sensitive blocking groups by the acidic ion-exchanger. The resin bed is washed with organic solvents and the blocked intermediate eluted with pyridine in dimethylformamide. Concentration and precipitation with ether eliminates the co-eluted 3-bromopyridine. The protected peptide intermediate is analyzed, characterized and, if this seems to be necessary, further purified. A second cycle of deprotection, coupling and

---

[1] The use of a mixture of the trifluoroacetate salt and of a tertiary amine instead of the amine-component itself is a simplification that can not be fully approved. In addition to some shortcomings discussed in Chapter VII it leads to partial trifluoroacetylation of the amino group because activation of trifluoroacetic acid by carbodiimides can not be avoided. Addition of 1-hydroxybenzotriazole does not completely eliminate this side reaction.

absorption follows and so on. While the picolyl ester method is adaptable also for segment condensation, it is best suited for the stepwise strategy.

At the final stage the ester group is cleaved by reductive methods such as catalytic hydrogenation, reduction with sodium in liquid ammonia or electrolytic reduction at a mercury cathode. Saponification with cold alkali is also feasible but can be recommended only with some reservation: it involves some risk of racemization and other side reactions.

The picolyl ester method was shown in several examples, including the preparation of fairly complex molecules, to be highly effective. Still, it is not widely used, probably because the process is not readily amenable to mechanization and automation. Also, it is time consuming and requires careful execution. Yet, in the preparation of peptides needed in high purity, this price might be worth paying.

# C. Rapid Synthesis

## 1. Chain-building Without Isolation of Intermediates

Facilitation of peptide synthesis by omitting isolation of blocked intermediates has often been proposed. Various methods of coupling were adopted for this purpose, for instance the azide procedure, mixed anhydrides and carbodiimides. The common feature of these proposals is extraction of unreacted starting materials and water soluble by-products from the reaction mixture with dilute HCl, bicarbonate solution and water. The organic layer should contain only the desired intermediate which is then used without isolation or further purification in the next deprotection-coupling cycle. Through the years several authors demonstrated the practicality of such simple procedures but none of these variations was received with the enthusiasm shown toward solid phase peptide synthesis. This reservation might be due to the circumstance that in these rapid methods examination of the intermediates and elimination of by-products not removed by washing is sacrificed without the principal gain associated with the solid phase method, namely easy, mechanized and automated execution.

## 2. Chain-building Through Isolated Intermediates

The efficient activation achieved with pentafluorophenyl esters allows the stepwise synthesis of peptides in remarkably short time. The crude acylation product, a blocked intermediate, is precipitated and washed, immediately deblocked and used as amine component in the incorporation of the next residue. A protected octapeptide with the sequence of angiotensin could be assembled in a single day (Kisfaludy and Nyéki 1975) and similarly fast chain-lengthening yielded the blocked form of oxytocin (Kisfaludy and Schön 1975).

# D. Synthesis "in situ"

A technique in which the intermediates are isolated but remain in the same vessel throughout the chain-lengthening process is feasible through the application of a centrifuge tube provided with a standard-taper joint (Bodanszky et al. 1973). This can be attached to a rotary evaporator, hence it is suitable both for the removal of solvents and for the separation of precipitates by centrifugation. A blocked intermediate, for instance a tert.butyloxycarbonyl tripeptide amide, is placed in the modified centrifuge tube and treated with an acidolytic reagent such as trifluoroacetic acid. Evaporation of the excess acid *in vacuo* is followed by precipitation with ether. The trifluoroacetate salt is dried, weighed, examined (by thin layer chromatography etc.) then dissolved in the solvent selected for coupling, e.g. dimethylformamide. A tertiary amine is added followed by the active ester of the blocked amino acid to be incorporated and — if desirable — by the addition of a catalyst, for instance 1-hydroxybenzotriazole. Completion of the acylation is monitored by spot tests with ninhydrin. Removal of the solvent *in vacuo,* precipitation with a "non-solvent" such as ether or ethyl acetate concludes the cycle. A suitably selected "non-solvent" not merely precipitates the blocked intermediate but also eliminates the excess active ester, the catalyst and the salt of the tertiary amine. The blocked peptide is washed, dried, weighed and examined. If necessary it can be recrystallized or reprecipitated, still in the same vessel. It is ready for the next cycle of stepwise chain-building. Fairly long peptides could be prepared in this maner, but the *"in situ"* technique has not been generally accepted so far.

# E. Simultaneous Synthesis of Peptide Analogs

The "in situ" technique, described in the last section, could be applied for the simultaneous synthesis of secretin and three of its analogs (Fink and Bodanszky 1976) simply by carrying out the process in four centrifuge tubes. In all chain-lengthening steps identical operations were performed on the intermediates in the four tubes except at the stages where an amino acid different from the residue in the parent sequence had to be introduced. Solid phase peptide synthesis is even more suitable for simultaneous chain building. In an interesting version of this method two resins were applied, polymers with sufficiently different densities to allow their separation through sedimentation in a medium with intermediate density. The two polymers, with the peptides attached to them, were handled as a mixture but were separated prior to steps in which a different residue had to be incorporated in each of them. Of course, they were again separated before the peptides were cleaved from their polymeric supports.

Further simplification in simultaneous chain-building can be achieved by placing small batches of an aminoacyl resin or peptidyl-polymer in individual containers which are permeable by the solvent. For instance tea-bags were proposed for this purpose. The assembly of several such aliquots is treated in the same way when identical residues are introduced and an appropriately tagged aliquot is picked out from the group when a change from the parent sequence has to be made. The possibility to prepare a large series of analogs of a biologically active peptide for screening in biological or pharmacological tests is certainly attractive. Yet, the facility of the process raises a certain concern about the quality of the products. Analysis and characterization must keep up with the proliferation of samples.

## F. Insoluble Reagents

### 1. Active Esters

Polymeric active esters represent an interesting counterpart to solid phase peptide synthesis. In the latter the amine-component is linked to an insoluble polymer while in the insoluble active esters the carboxyl-component is polymer bound. The main advantage of these acylating agents lies in the facile removal, by filtration, of the leaving by-product together with the unreacted excess of the reagent. The unreacted portion of the insoluble active esters can be utilized for the incorporation of the same residue at another stage or in another synthesis.

The earliest examples of insoluble active esters were derivatives of phenols provided with an ion-forming group by which they can be attached to an ion-exchange resin (Wieland and Birr 1966):

Alternative realizations of this novel approach are nitrated polymeric phenols (Fridkin et al. 1968) and a polymeric form of N-hydroxysuccinimide (Laufer et al. 1968):

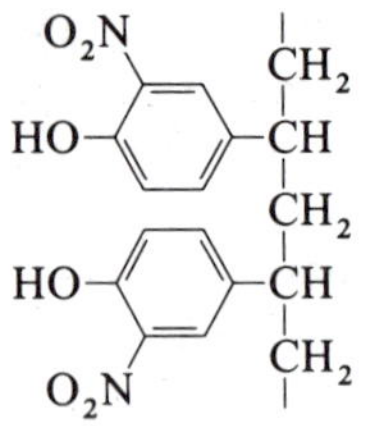

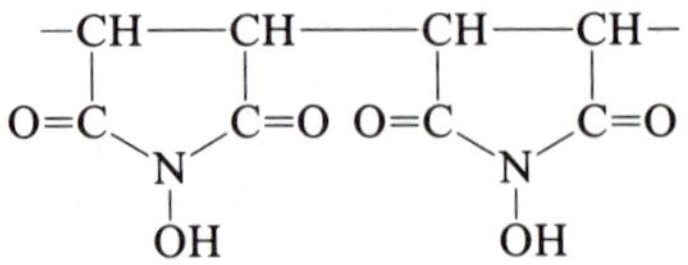

Esterification of the hydroxyl group in these materials with a protected amino acid is carried out in the usual manner through activation of the carboxyl group with a carbodiimide. The resulting reactive derivatives are aminolyzed by the amine-component

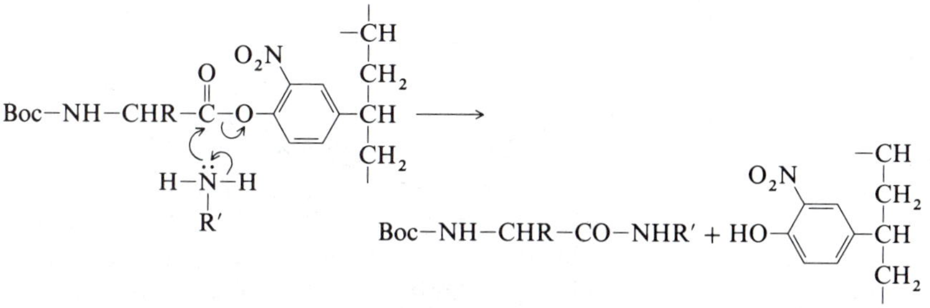

$$Boc-NH-CHR-\overset{O}{\underset{\|}{C}}-O-\underset{O_2N}{\bigcirc}\text{-CH(-CH}_2\text{-CH-CH}_2\text{-)}$$

$$H-\overset{..}{N}-H \quad \underset{R'}{\big|}$$

$$\longrightarrow \quad Boc-NH-CHR-CO-NHR' + HO-\underset{O_2N}{\bigcirc}\text{-CH(-CH}_2\text{-CH-CH}_2\text{-)}$$

dissolved in an organic solvent in which the polymeric reactant shows good swelling. The active ester content of the insoluble material can be assayed before and after acylation, for instance, through the reaction of a small sample with benzylamine. The unreacted portion of the base is determined by non-aqueous titration with a strong acid.

A word of caution may be necessary here. This ingenious approach is readily executed as long as the amine-component is indeed soluble in the selected solvent. The often encountered difficulty, formation of poorly soluble blocked intermediates, can produce viscous gels from which separation of the particles of the polymeric acylating agent, might turn out to be impractical.

Further development led to improved polymeric active esters in which a spacer is placed between the activating group and the polymer, as in

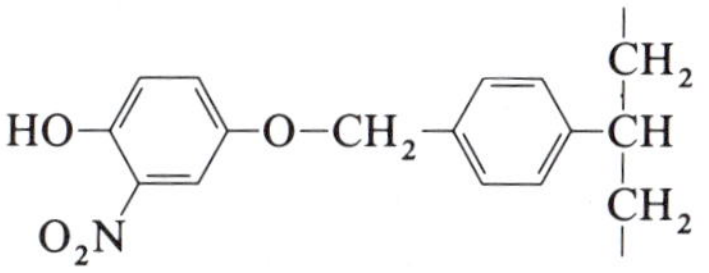

$$HO-\underset{O_2N}{\bigcirc}-O-CH_2-\bigcirc-CH(-CH_2-CH_2-CH-)$$

or in the activated amino acid derivative attached to polyethyleneglycol:

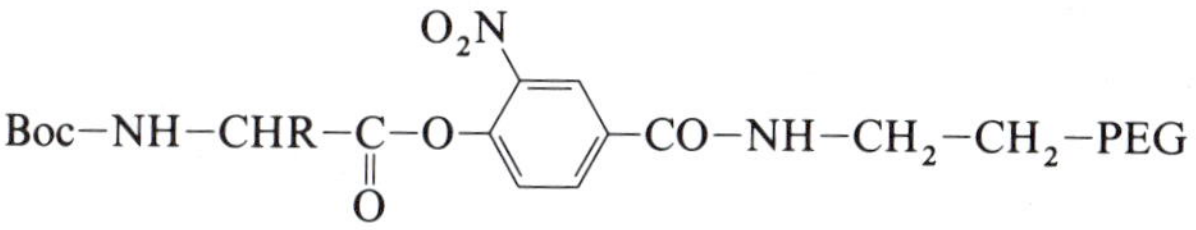

$$Boc-NH-CHR-\underset{O}{\overset{\|}{C}}-O-\underset{O_2N}{\bigcirc}-CO-NH-CH_2-CH_2-PEG$$

Similarly a spacer was inserted into polymeric analogs of 1-hydroxybenzotriazole esters:

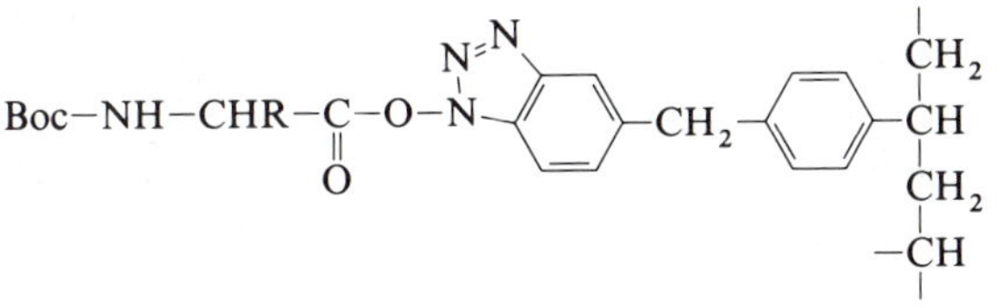

$$Boc-NH-CHR-\underset{O}{\overset{\|}{C}}-O-N\underset{N\overset{N}{\diagdown}}{\bigcirc}-CH_2-\bigcirc-CH(-CH_2-CH-CH_2-CH-)$$

Insoluble active esters open up new possibilities. Thus, the solution of the amine-component can be passed through a column of such an acylating agent and the acylation product will be in the effluent. After appropriate deblocking the new amine-component is passed through a second column containing the next amino acid to be incorporated, of course in protected and activated form. For the full realization of this principle in a continuous process it will be necessary to find temporary blocking groups that produce only inert materials. Otherwise the accumulation of by-products formed in most methods of deprotection, such as tert.butyl trifluoroacetate or trialkylammonium salts produced in the treatment of amine-salts with tertiary bases, will interfere with the procedure.

An already practical exploitation of the insoluble active ester concept is their use in the synthesis of cyclic peptides. The deprotected amino group at the N-terminus of a chain of suitable length attacks the reactive ester carbonyl at the C-terminus:

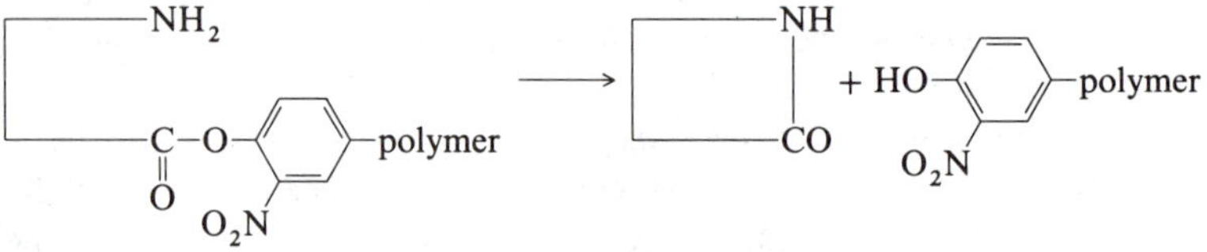

No or little ring closure can be expected in chains containing less than six amino acid residues. For better control it appears to be preferable to apply esters which are fairly inactive but can be activated when cyclization is desired. For instance, oxidation of a thioether to the electron-withdrawing sulfone

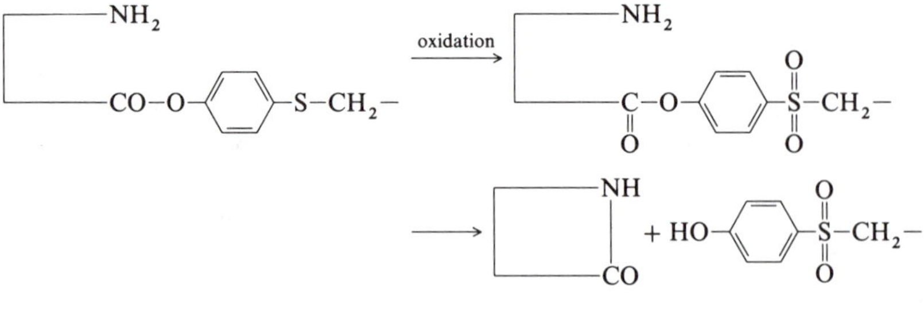

can provide the needed increase in reactivity.

Finally, while it is not obvious how both the amine-component *and* the carboxyl component could be applied in polymeric form, this dilemma was solved (Cohen et al. 1981) by the introduction of a mediator. A solution of imidazole is circulated through separate beds of the insoluble reactive carboxyl-component and the polymer-bound amine-component. The acyl-imidazoles generated in the active ester bed are excellent acylating agents which transfer the acyl group to the nucleophile in the other bed and thus convert the amine-component to the desired, still polymer bound, blocked intermediate.

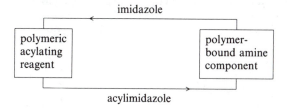

imidazole

| polymeric acylating reagent | | polymer-bound amine component |

acylimidazole

## 2. Polymeric Coupling Reagents

These attractive but somewhat neglected tools might become more frequently used in the future. An early example, polyhexamethylenecarbodiimide (Wolman et al. 1967)

$$-(CH_2)_6-N=C=N-(CH_2)_6-N=C=N-(CH_2)_6-N=C=N-(CH_2)_6-N=C=N-$$

was followed by several other implementations of this idea, of which the polymer-bound form of 1-ethoxycarbonyl-2-ethoxy-1,2-dihydroquinoline (EEDQ)

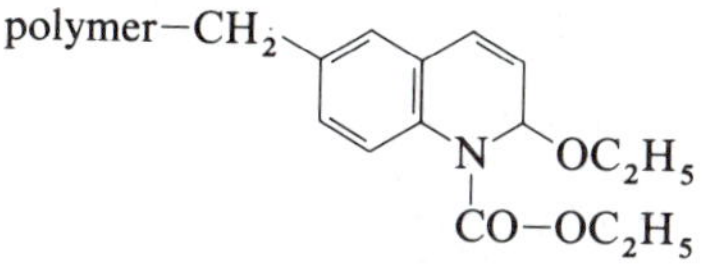

polymer$-CH_2$

N $OC_2H_5$
$CO-OC_2H_5$

can be mentioned.

## 3. Insoluble Cleaving Reagents

Removal of the *o*-nitrobenzenesulfonyl group with a polymeric form of 2-mercaptopyridine

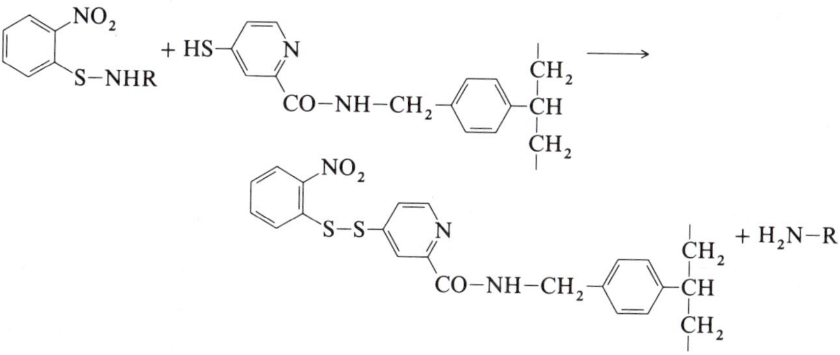

$NO_2$

$S-NHR$  $+ HS-$N

$CO-NH-CH_2-$ $CH_2$ $CH$ $CH_2$

$NO_2$

$S-S-$N

$CO-NH-CH_2-$ $CH_2$ $CH$ $CH_2$  $+ H_2N-R$

has an obvious advantage over the simple, soluble thiol-reagent: the cleavage product is trapped on the polymer and removed from the reaction mixture by filtration. In an analogous manner the by-product generated in the removal of the 9-fluorenylmethyloxycarbonyl (Fmoc) group, dibenzofulvene, is bonded to the insoluble reagent if a polymer-bound secondary amine is applied for cleavage:

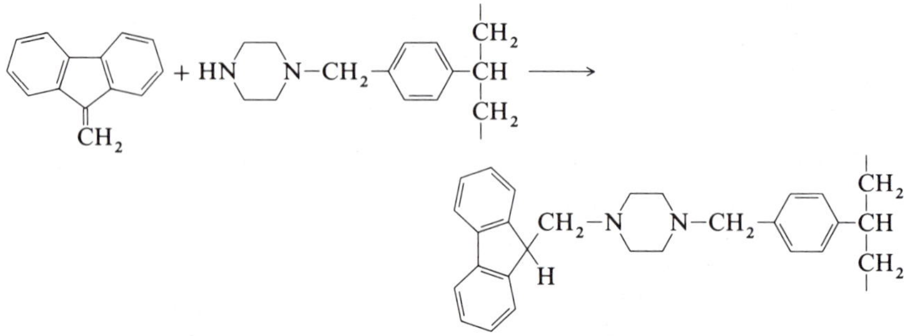

# References

Bodanszky, M., Funk, K. W., Fink, M. L.: J. Org. Chem. *38*, 3565 (1973)
Camble, R., Garner, R., Young, G. T.: Nature *214*, 247; J. Chem. Soc. *1969*, 1918
Cohen, B. J., Kraus, M. A., Patchornik, A.: J. Amer. Chem. Soc. *103*, 762 (1981)
Fink, M. L., Bodanszky, M.: J. Amer. Chem. Soc. *98*, 974 (1976)
Fridkin, M., Patchornik, A., Katchalski, E.: J. Amer. Chem. Soc. *88*, 3164 (1966)
Laufer, D. A., Chapman, T. M., Marlborough, D. I., Vaidya, V. M., Blout, E. R.: J. Amer. Chem. Soc. *90*, 2696 (1968)
Kisfaludy, L., Nyéki, O.: Acta Chim. Acad. Sci. Hung. *86*, 343 (1977)
Kisfaludy, L., Schön, I.: Acta Chim. Acad. Sci. Hung. *84*, 227 (1975)
Shemyakin, M. M., Ovchinnikov, Y. A., Kiryushkin, A. A., Kozhevnikova, I. V.: Tetrahedron letters *1965*, 2323
Wieland, T., Birr, C.: Angew. Chem. Int. Ed. *5*, 310 (1966)
Wolman, Y., Kivity, S., Frankel, M.: Chem. Commun. *1967*, 629

# Additional Sources

Fridkin, M.: Polymeric Reagents in Peptide Synthesis, in The Peptides, vol. 2, Gross, E., Meienhofer, J. eds. pp. 333–363, New York, Academic Press 1980
Kisfaludy, L.: Repetitive Methods in Solution, *ibid.* pp. 417–440
Mutter, M., Bayer, E.: The Liquid-Phase Method for Peptide Synthesis, *ibid.* pp. 285–332

# XII. Analysis and Characterization of Synthetic Peptides

The molecular weight of most synthetic organic compounds is well below 1000 daltons. The products of peptide synthesis, however, have frequently molecular weights that exceed this number. Calculating with an average residue weight of 115 even a nonapeptide's molecular weight is more than 1000 and a further increase is caused by the acids (e.g. acetic acid) associated with cations in the side chain of basic residues. In protected intermediates the blocking groups contribute in a major way to the size of the molecule. While high molecular weight in itself could be the cause of some difficulties in analysis, the problem is compounded by the similar elemental composition of the amino acid constituents. Thus, incorporation of a residue leads to only a minor change in the values of elemental analysis and in the physical properties of intermediates. Furthermore, the by-products formed in various operations of synthesis are often quite similar in their composition to the desired peptide derivatives. These complexities obviously increase the need for thorough scrutiny of synthetic peptides, if possible by a battery of tests. In the following paragraphs we will attempt to point out some well established methods of analysis that are useful in this area.

## A. Homogeneity Tests

First and foremost, synthetic peptides should be tested for homogeneity. It is time consuming and mostly unrewarding to analyze the amino acid composition or elemental composition of mixtures. Similarly, the results of spectroscopic examinations are fully meaningful only if homogeneous samples are studied. Sharp melting point is an encouraging indication but not the final word on purity. Yet, even the simplest and least expensive method, thin layer chromatography (tlc) can detect the most common impurities, unreacted starting materials, and also the majority of by-products. Of course, interpretation of chromatograms must be based on experience and judgement. The welcome observation of a single spot can be misleading: in some solvent systems the affinity of one of the solvent components to silica gel might cause the formation of a second solvent front to which several materials can travel and appear in one line. Furthermore, as it often happens, the method of detection fails to reveal certain materials. Most color reagents are too specific to detect *all* substances and even in the least specific process, charring, thermally stable and volatile compounds, such as diketopiperazines or the ubiquitous N,N'-dicyclo-hexylurea (DCU), can escape and be overlooked. The fairly general method of chlorination of the amide groups followed by a color reaction based on the oxidizing effect of chlorimides, is not absolutely reliable: in some cyclic peptides the amide groups are not accessible for the commonly used chlorinating re-agent, butyl hypochlorite. It might be possible to remedy such situations, for instance by the use of chlorine gas, but, in general, it is advisable to apply more than one method of detection. Exposure to iodine vapors could be one of them.

The opposite complication, two spots produced by a single substance, can also occur. Peptide acetates, trifluoroacetates, etc., are transformed, in part, to the corresponding hydrochlorides by hydrochloric acid in the silica gel. In *p*-toluenesulfonate salts displacement by hydrochloric acid results in the ap-pearance of an additional spot, that of *p*-toluenesulfonic acid. Such difficulties must be reduced and misleading coincidences, such as two peptides with closely similar Rf values avoided by the application of two or more solvent systems. The composition of these thin layer chromatography systems should be sub-stantially different from each other, not merely the same solvents used in modified ratios. The most frequently applied systems, methanol-chloroform, n-butanol-acetic acid-water and n-butanol-pyridine-acetic acid-water, are gen-erally satisfactory, but if problems arise, they should be supplemented by additional solvent systems. For instance replacement of butanol by ethyl ace-tate provides a less viscous and hence faster moving solvent system with high resolving power.

The advent of high pressure liquid chromatography (hplc) revolutionized the examination of peptides and the commercially available reverse-phase columns allow rapid separation, detection and quantitation of the components in a mixture and also automatic recording of the results. No wonder that the method gained enormous popularity in short time. Enthusiastic users of hplc sometimes forget, however, that a certain amount of reservation is needed even with the best methods. Thus, materials purified by preparative hplc are often analyzed by hplc in the same system. This is obviously incorrect and the results can be quite misleading. Also, too much trust is placed in the quantitative analysis of mixtures although the numbers in the print-out are merely integra-tion values of areas under the peaks recorded according to absorption in the ultraviolet. The tacit, but mostly unjustified, assumption was made by the manufacturer, that at the recording wavelength the various components of a mixture have the same specific extinction coefficient. Therefore, small amounts of impurities with extinction coefficients that are orders of magnitude higher than the extinction coefficient of the main component can make materials approaching homogeneity appear as grossly impure. The opposite is equally possible.

Free peptides can be examined by paper electrophoresis or by thin layer electrophoresis as well. If one or more cation forming groups, (amino group, guanidino group, imidazole) are present, electrophoresis in an acidic solvent, such as dilute acetic acid, is most revealing. Compounds with free carboxyl groups are best run under basic conditions to allow differentiation by the number of carboxylate anions. The acidic character of the phenolic hydroxyl in tyrosine should be included in these considerations.

If any of these methods indicates more than trace amounts of impurities then the synthetic peptide must be further purified prior to more extensive analysis or characterization. Chromatography and countercurrent distribution are the most commonly applied methods for the purification of peptides and of blocked intermediates, but the classical procedure of crystallization remains the simplest and sometimes most effective approach.

## B. Amino Acid Analysis and Sequence Determination

Determination of amino acid composition is, beyond doubt, the most informative method in the analysis of synthetic peptides. The preceding hydrolysis of peptide bonds must of course be complete. This can be achieved by dissolving the sample in constant boiling hydrochloric acid and heating of the solution, preferably in an evacuated and sealed ampoule, at 110 °C for 16 hours. Some blocked intermediates, however, are insoluble in the ca 6 N HCl even at its boiling point. In such cases mixtures of HCl with formic acid or acetic acid can solve the problem. For the hydrolysis of polymer-bound peptides often an HCl-propionic acid mixture is applied. The presence of methionine sulfoxide in synthetic peptides deserves special consideration. On hydrolysis with HCl both methionine and its sulfoxide appear on the recordings of the amino acid analyzer, but not necessarily in the ratio they occur in the sample. The sulfoxide is partially reduced:

$$
\begin{array}{ccc}
\begin{array}{l}
CH_3 \\
| \\
S=O \\
| \\
CH_2 \\
| \\
CH_2 \\
| \\
H_2N-CH-COOH
\end{array}
&
+\ 2\,HCl \rightleftharpoons
&
\begin{array}{l}
CH_3 \\
| \\
S \\
| \\
CH_2 \\
| \\
CH_2 \\
| \\
H_2N-CH-COOH
\end{array}
\quad +\ Cl_2 + H_2O
\end{array}
$$

Since the chlorine generated in the process might be consumed, for instance by the chlorination of the phenolic ring of tyrosine residues, it is worthwhile to add a small amount of a reducing agent, such as $\beta$-mercaptoethanol to the hydrolysis mixture. In this case only methionine will be found in the hydrolysate. If information on the methionine sulfoxide content of the peptide is also needed, then instead of HCl rather an aqueous solution of methanesulfonic acid should be used for hydrolysis: this leaves the sulfoxide intact.

For complete hydrolysis of bonds between hindered amino acid residues the above mentioned 16 hour period is not sufficient. In addition to isoleucine and valine synthetic intermediates can contain residues with bulky and acid resistant blocking groups, such as S-benzylcystein. Should the latter be preceded or followed by a hindered amino acid, then the time of hydrolysis must be considerably prolonged. Since the hydroxyamino acids and also tryptophan gradually decompose under these conditions, it might be necessary to carry out more than one analysis, with shorter and longer hydrolysis times.

In the evaluation of the results of analyses of segment condensation products special attention is given to amino acids that occur in only one of the peptides coupled in the reaction. If the carboxyl component contains residue A which is absent from the amine component, while the opposite is true for residue B, then the presence of equimolar amounts of amino acids A and B in the hydrolysate is usually regarded as evidence for successful coupling. Of course, the conclusion drawn from the analysis of "diagnostic" residues is valid only in conjunction with evidence for homogeneity of the analysed sample. For instance, the coupling of segments by the azide method (cf. Chapter V) is often accompanied by Curtius rearrangement of the azide resulting in a condensation product which is different from the desired material but can have "diagnostic" residues in the proper molecular ratio. On the other hand, the *absence* from the hydrolysate of the amino acid which was activated in the form of acid azide is evidence for Curtius rearrangement:

$$-NH-CHR-CO-NH-CHR'-\overset{\overset{\displaystyle O}{\|}}{C}-N=\overset{+}{N}=\overset{-}{N} \xrightarrow{-N_2}$$

$$-NH-CHR-CO-NH-CHR'-N=C=O$$

$$-NH-CHR-CO-NH-CHR'-N=C=O + H_2N-CHR''-CO- \longrightarrow$$

$$-NH-CHR-CO-NH-CHR'-NH-CO-NH-CHR''-CO- \xrightarrow{HCl/HOH}$$

$$H_2N-CHR-COOH + 2NH_3 + OCHR' + CO_2 + H_2N-CHR''-COOH$$

Last but not least the *recovery* of amino acids in the hydrolysate, that is the molar amounts found versus the aliquot of the sample hydrolyzed, is a good measure of the peptide content of a synthetic sample. The value of this information should not be belittled. Samples of synthetic peptides more often than not contain moisture, organic solvents and sometimes even inorganic impurities. Also, it is not unusual that the amount of acetic acid found by analysis of acetate salts exceeds the amount calculated from the number of basic centers in the molecule. Therefore, the weight of a sample does not reveal its peptide content and quantitative amino acid analysis starting with the hydrolysis of an exactly weighed amount of the peptide is a simple and reliable method for its determination.

In laboratories where an automatic sequencing instrument is routinely used, it is worthwhile to apply it for the examination of synthetic materials, or at least of the final product of a synthesis. Shortcomings of the chain-building process, such as deletion of a residue (sometimes encountered in solid phase peptide synthesis) are uniquely recognized through sequencing.

# C. Elemental Analysis

Because the values determined by elemental analysis are not grossly different from peptide to peptide, the information gained in this traditional method is often deemed not significant enough to warrant the sacrifice on time and material. Experience, however, shows that while correct values of elemental composition determined by combustion are indeed not sufficient to prove the purity of a synthetic peptide, values which are *not* in agreement with the calculated ones do alert the investigator and prompt a more detailed examination which should then detect the reasons underlying the discrepancy.

A commonly encountered problem in elemental analysis is finding low values both for carbon and nitrogen. A concurrently found high value for hydrogen indicates moisture in the sample. Freeze-dried materials can contain as much as 50% water which is not removed by drying except if this is done at elevated temperature and for prolonged period of time. The same is true for the elimination of solvents retained in blocked peptides. Both protected and free peptides can behave like hygroscopic materials, in the sense known for cotton. This means, that dried samples absorb moisture from the air without being deliquescent. They must be weighed, therefore, in closed vessels after drying, and handled with the exclusion of moisture.

Low values for carbon and nitrogen, but with correct C/N ratio and without a high value for hydrogen, suggest the presence of inorganic impurities. A final purification by gel-permeation chromatography, for instance on sephadex columns, is generally sufficient for the removal of such impurities, but it should be remembered that peptides are good ligands for metal ions and the complexes are often quite stable. Frequently encountered impurities, like acetic acid or trifluoroacetic acid in excess over the amount neutralizing the basic groups are recognized by comprehensive analysis that extends to all elements present. An extra benefit of such analysis is the built-in control: the results must add up to 100%. In some problematic situations determination of functional groups, acid-base titrations, etc. is advisable. The possible presence of chloride, fluoride or trifluoromethanesulfonate ions, stemming from the acid employed in deprotection, should not be overlooked.

## D. Measurement of Optical Activity

Determination of specific rotation in itself provides no significant information about a synthetic material. Also, the frequently made assumption that among different preparations of the same peptide the one with highest value of specific rotation has the highest chiral purity as well, can be wrong. For instance, dehydration of the asparagine side chain yielding a $\beta$-cyanoalanine residue is accompanied by a considerable increase in rotatory power and this increase might compensate a decrease caused by racemization. In spite of these reservations it would be a mistake to forgo the determination of optical activity. Specific rotation is a characteristic physical constant, usually different for different peptides and discrepancies between reported and found values should prompt additional investigations. It is particularly important to compare the specific rotation of a synthetic peptide with that of the corresponding natural product. Of course, when the purpose of the synthesis is independent evidence for the correctness of a proposed structure, this comparison becomes indispensable.

In the determination of optical activity it is generally advisable to use conditions (temperature, concentration, solvent) that are not grossly different from the ones reported in the literature. In aqueous solutions the pH has major influence on the results. Specific rotation of peptides can drastically change on the addition of acids and bases. The value determined for a sample of lysine vasopressin acetate underwent a substantial change when it was dried at elevated temperature. This seemingly mysterious effect was merely the consequence of the loss of one equivalent of acetic acid on drying. The discrepancy disappeared when dilute acetic acid rather than water was used as solvent in the determination. Problems and contradictions encountered in the determination of specific rotation can sometimes be resolved by extending the examination from reading at a single wavelength to the recording of the optical rotatory dispersion spectrum.

## E. Infrared and Ultraviolet Absorption Spectra

Infrared spectra are valuable in the examination of small molecular-weight intermediates, such as blocked and activated derivatives of amino acids. The stretching frequency for the urethane carbonyl (around $1700 \text{ cm}^{-1}$) is distinguishable from that of the activated carbonyl group in anhydrides (over $1800 \text{ cm}^{-1}$) or in active esters (around $1800 \text{ cm}^{-1}$). Also, changes in ir spectra were used for monitoring transformations in polymeric supports. For instance, conversion of chloromethylated copoly-styrene-divinylbenzene to the acetoxy derivative can be followed by the appearance and growth of an ester carbonyl band while the following saponification by the gradual disappearance of the same absorption band.

Less revealing are the ir spectra of larger peptide molecules. Unless sophisticated instruments, such as high resolution Fourier-transform spectrophotometers are applied, the large amide absorption overshadows other carbonyl frequencies. In some special cases, however, ir spectra can be uniquely informative. For example esterification of the phenolic hydroxyl in tyrosine with sulfuric acid is best revealed through the characteristic ir spectrum of sulfate esters, clearly distinguishable from the spectrum of the sulfonic acid derivatives which are the potential by-products of the sulfation process.

Ultraviolet spectra provide excellent information about tyrosine and tryptophan residues even in large peptides. In neutral solution the maxima of the main absorption bands for the aromatic chromophores in the two amino acids are too close to each other to allow clear distinction between them. The phenol in the side chain of tyrosine is readily changed, however, to phenolate by the addition of alkali or an organic base. This results in a 20 nanometer shift of the absorption maximum toward the visible. Hence, it is possible to determine the amounts of both tyrosine and tryptophan residues in peptides merely by the examination of uv spectra. Other chromophores in amino acid side chains are too weak for simple studies, but in the absence of tyrosine and tryptophan the absorption of phenylalanine side chains can be determined and in some peptides the absorption of disulfide bridges as well. Some blocking groups are more important contributors to absorption in the uv. Thus, the nitro group in nitroarginine residues or the fluorenyl system in Fmoc-derivatives are powerful chromophores and certainly should not be overlooked in the examination of peptide intermediates by uv spectrophotometry.

# F. Nuclear Magnetic Resonance Spectra

The value of high resolution nuclear magnetic resonance (nmr) spectroscopy and of two dimensional nmr spectra has been stressed in Chapter IV. The effort needed for the recording and interpretation of these spectra is not necessarily justified in the examination of synthetic materials. The information, however, that can be gained about synthetic intermediates and final products by proton nmr spectroscopy with simple instruments and in short time should not be underestimated. Certain blocked peptides particularly lend themselves to analysis via nmr spectra. The aromatic resonances corresponding to benzyl groups and benzyloxycarbonyl groups when compared with the conspicuous peaks (9-proton singlet) of tert.butyl groups can furnish evidence for the success of a coupling reaction. Well separated aromatic peaks representing protons in the imidazole ring of histidine or in the indole nucleus of tryptophan are identified without difficulty. It is less simple to assign individual protons to resonances on OH or NH groups or the peaks of hydrogens bound to $\alpha$- and $\beta$-carbon atoms of amino acid residues. An exception is presented by the methyl protons of alanine: the methyl group being close to the electron-withdrawing environment of the peptide backbone is readily recognized as a doublet shifted downfield from other methyl resonances. It might not always be simple to analyze the various other methyl protons in the spectrum of larger peptides, but at least their sum can be determined by integration of the composite peak and this value compared with that of a well defined resonance. In methionine containing peptides the S-methyl group is particularly suitable for such comparison.

A frequently encountered problem in the recording of nmr spectra of blocked peptides is (unless solid state spectra are sought) to find a suitable solvent. Although $CDCl_3$ is applicable for small protected intermediates, the majority of the samples is not sufficiently soluble in deuterio-chloroform. Deuterated dimethylformamide and dimethylsulfoxide are more auspicious, but the chemical shifts are rather different from those observed in $CDCl_3$. This is even more so in trifluoroacetic acid, an excellent solvent for most intermediates but obviously not inert toward them. A good compromise can be found in the use of fully deuterated acetic acid, $CD_3COOD$. The chemical shifts found in this solvent are quite close to those recorded in $CDCl_3$ and the marked simplification of the spectra due to the exchange of amide and hydroxyl protons by the ionizable deuterium atom of the solvent is an added benefit. If a comparison with a spectrum in which no exchange took place is desirable, $CD_3COOH$ can be applied.

Deprotected peptides are usually soluble in water, hence $D_2O$ is a practical solvent for recording of their nmr spectra. If necessary, consideration can be given to $CD_3COOD$. If a sample is insoluble both in water and in acetic acid, then a mixture of $D_2O$ and $CD_3COOH$ might provide a way out of the dilemma. The somewhat disturbing peak of DHO, due to the partial exchange of protons by $D_2O$, can be diminished or eliminated by driving the exchange to completion through evaporation and redissolution of the sample in fresh solvent. An important piece of information from nmr spectra of final products is the absence of resonances which correspond to protecting groups. Several instances have been reported in the literature where certain blocking groups could not be removed under the usually applied conditions. A benzyl group might be tenaciously retained on the hydroxyl of a threonine residue or on the imidazole nitrogen of a histidine side chain. Such imperfections appear with great clarity in proton nmr spectra of free peptides. Needless to say that additional useful information can be gained with the help of carbon-13 spectra or by nmr spectroscopy based on other nuclei, but proton nmr spectra are generally sufficient for the scrutiny of most intermediates and final products of peptide synthesis.

# G. Molecular-weight Determination

The molecular weight of synthetic peptides is usually beyond question. Notable exceptions are cyclic peptides because in the process of ring-closure instead of simple cyclization cyclodimerization can occur. The reasons for this well studied phenomenon were discussed in connection with synthesis of cyclic peptides (Chapter IX). Molecular-weight determination with the help of mass spectra seems to be a simple way to answer questions about the occurrence or absence of cyclodimerization, but the results could turn out less unequivocal than desired. The smaller molecular-weight product of simple cyclization might be present merely as an impurity, yet on account of its higher volatility and better thermal stability it can appear as the principal component. Because of self-association of peptide molecules both gel-filtration and sedimentation in the ultracentrifuge might not be entirely reliable for the determination of molecular weight of peptides. Methods based on particle-number (melting point depression, boiling point elevation, osmotic pressure) can produce similarly misleading results with peptides that have a tendency for aggregation. The situation is further complicated in compounds with ionizable groups such as free carboxyls, amino or guanidino groups. Whether or not the counter-ions appear as separate particles depends on dissociation which, in turn, is a function of the solvent applied in the study. These kind of complications are avoided in a "chemical" approach, molecular-weight determination by the method of least substitution. Its value was demonstrated in investigations of gramicidin S (Battersby and Craig 1951). Treatment with less than one equivalent of 2,4-dinitrofluorobenzene yielded a mixture of unsubstituted, monosubstituted and disubstituted peptides. The dinitrophenyl (DNP) derivatives were separated and the uv absorption of the monosubstituted compound determined. Its molecular weight was then calculated with the help of the known molar absorption coefficient of the 2,4-dinitrophenyl group:

$$
\begin{array}{l}
\left[\begin{array}{l}
\text{Val}-\text{Orn}-\text{Leu}-\text{D}-\text{Phe}-\text{Pro} \\
\\
\text{Pro}-\text{D}-\text{Phe}-\text{Leu}-\text{Orn}-\text{Val}
\end{array}\right]
\end{array}
+ \text{F}-\text{C}_6\text{H}_3(\text{NO}_2)_2 \text{ (less than 1 equiv.)} \xrightarrow{\text{NaHCO}_3}
$$

$$
\left[\begin{array}{l}
\overset{\text{DNP}}{\text{Val}}-\text{Orn}-\text{Leu}-\text{D}-\text{Phe}-\text{Pro} \\
\\
\text{Pro}-\text{D}-\text{Phe}-\text{Leu}-\underset{\text{DNP}}{\text{Orn}}-\text{Val}
\end{array}\right]
+
\left[\begin{array}{l}
\overset{\text{DNP}}{\text{Val}}-\text{Orn}-\text{Leu}-\text{D}-\text{Phe}-\text{Pro} \\
\\
\text{Pro}-\text{D}-\text{Phe}-\text{Leu}-\text{Orn}-\text{Val}
\end{array}\right]
\begin{array}{l}
\text{unchanged} \\
+ \text{ starting} \\
\text{material}
\end{array}
$$

With appropriate modifications the method is applicable for any peptide that can be tagged with a chromophore.

# Reference

Battersby, A. R., Craig, L. C.: J. Amer. Chem. Soc. *73*, 1887 (1951); cf. also *ibid. 74*, 4023 (1952)

# Additional Sources

Benson, J. R., Louie, P. C., Bradshaw, R. A.: Amino Acid Analysis of Peptides, in: The Peptides, Vol. 4, Gross, E., Meienhofer, J., eds. pp. 217–260, New York, Academic Press 1981

Stein, S.: Ultramicroanalysis of Peptides and Proteins by High-performance Liquid Chromatography and Fluorescence Detection, in: The Peptides, Vol. 4, Gross, E., Meienhofer, J., eds. pp. 261–283. New York, Academic Press 1981

# Subject Index

H.-G. Elias

# Mega Molecules

Tales of Adhesives, Bread, Diamonds, Eggs, Fibers,
Foams, Gelatin, Leather, Meat, Plastics, Resists,
Rubber, . . . and Cabbages and Kings

1987. 55 figures, 34 tables. XIII, 202 pages.
ISBN 3-540-17541-5

**Contents:** Genuine Plastics and Other Natural
Products. – In the Beginning was the Deed. – How
Big is Big? – False Doctrines. – The Mysterious Crazy
Glue. – Corn Syrup and Hi-Tech. – Engine Oils and
Vanilla Sauces. – Screwing Up Things. – Spiders,
Weavers, and Webs. – How to Iron Correctly. – From
Cheap Substitutes to High Performance Materials. –
Everything Flows. – In and Out. – Charges and
Currents. – Suggested Readings. – Appendix. –
Subject Index.

All life is based on big molecules, scientifically called
"macromolecules". Humans, animals, and plants
cease to exist without these structural, reserve, and
transport molecules. No life can be propagated
without macromolecular DNA and RNA. Without
macromolecules, we would only dine on water,
sugars, fats, vitamins and salts but would have to
relinquish meat, eggs, cereals, vegetables, and fruits.
We would not live in houses since wood and many
stones consist of macromolcules. Without macromolecules, we would have no clothes since all fibers
are made from macromolecules. No present-day car
could run: all tyres are based on macromolecules.
Without macromolecules, no photographic films, no
electronics.
Therefore, this booklet wants to lead from the experience of daily life to the concept of the structure
and function of macromolecular compounds. Properties of glues, plastics, multigrade engine oils, rubbers,
foams, etc., will be traced back to their chemical and
physical structures. The hardening of modern glues
and the sweetening of old potatoes will be discussed
as examples of chemical reactions of macromolecules, the staling of bread and the ironing of fabrics as
examples of physical transitions.

Springer-Verlag
Berlin Heidelberg New York
London Paris Tokyo

Springer